Sustainable Finance

Series Editors

Karen Wendt, CEO. Eccos Impact GmbH, President of SwissFinTechLadies, President Sustainable-Finance, Cham, Switzerland

Margarethe Rammerstorfer, Professor for Energy Finance and Investments, Institute for Finance, Banking and Insurance WU Vienna, Vienna, Austria

Sustainable Finance is a concise and authoritative reference series linking research and practice. It provides reliable concepts and research findings in the ever growing field of sustainable investing and finance, SDG economics and Leadership with the declared commitment to present the theories, methods, tools and investment approaches that can fulfil the United Nations Sustainable Development Goals and the Paris Agreement COP 21/22 alongside with de-risking assets and creating triple purpose solutions that ensure the parity of profit, people and planet through choice architecture passion and performance. The series addresses market failure, systemic risk and reinvents portfolio theory, portfolio engineering as well as behavioural finance, financial mediation, product innovation, shared values, community building, business strategy and innovation, exponential tech and creation of social capital. Sustainable Finance and SDG Economics series helps to understand keynotes on international guidelines, guiding accounting and accountability principles, prototyping new developments in triple bottom line investing, cost benefit analysis, integrated financial first plus impact first concepts and impact measurement. Going beyond adjacent fields (like accounting, marketing, strategy, risk management) it integrates the concept of psychology, innovation, exponential tech, choice architecture, alternative economics, blue economy shared values, professions of the future, leadership, human and community development, team culture, impact, quantitative and qualitative measurement, Harvard Negotiation, mediation and complementary currency design using exponential tech and ledger technology. Books in the series contain latest findings from research, concepts for implementation, as well as best practices and case studies for the finance industry.

This book series is indexed in SCOPUS.

Bertaç Şakir Şahin · Barış C. Cantürk
Editors

Fintech and Sustainable Development

A Policy-Oriented Comparison of Regulations in the EU, Germany, and Emerging Markets

Springer

Editors
Bertaç Şakir Şahin
Department of Business Administration
Yıldız Technical University
Istanbul, Türkiye

Barış C. Cantürk
Faculty of Law
Marmara University
Istanbul, Türkiye

ISSN 2522-8285 ISSN 2522-8293 (electronic)
Sustainable Finance
ISBN 978-3-032-31019-4 ISBN 978-3-032-31020-0 (eBook)
https://doi.org/10.1007/978-3-032-31020-0

This Springer imprint is published by the registered company Springer Nature Switzerland AG
The registered company address is: Gewerbestrasse 11, 6330 Cham, Switzerland

Preface

Fintech and sustainable development play a profound role in the global financial ecosystem, which produces intertwined impacts and outcomes within the realm of law and regulation. For that reason, this study examines the subject from this perspective. In this regard, this book consists of eight chapters. Following the introduction, which provides a general overview of the topic and a literature review, the subsequent chapters address the subject from both financial and legal perspectives. Navigating these intersecting yet inherently distinct specific fields requires the collaborative contributions of expert academics and practitioners.

Therefore, this book is the culmination of over a year of intensive effort by the members of the Future Finance Law Hub's Working Group-1: Fintech and Law, each of whom possesses the aforementioned expertise. We extend our deepest gratitude to all the contributors who made this study possible. The Future Finance Law Hub is a policy-making hub that aims to generate unique, comprehensive, and implementable policies for the futuristic issues at the intersection of law, technology, and finance by bringing together stakeholders from diverse disciplines and sectors. The incubation of the Future Finance Law Hub was kindly supported through the Friedrich Schiedel Fellowship hosted at the TUM School of Social Sciences and Technology and the TUM Think Tank in Munich, Germany, between September 2024 and September 2025. We would like to express our sincere thanks once again for their generous support. Following this period, the Hub has continued its operations based in Istanbul, Türkiye.

We believe that this work will significantly contribute to both academia and the business, serving as a valuable resource for scholars, practitioners, and policymakers alike.

Istanbul, Türkiye
March 2026

Bertaç Şakir Şahin
Barış C. Cantürk

Contents

Introduction 1
Bertaç Şakir Şahin

Fintech Investment Trends & Existing Studies on Fintech Regulations and Sustainable Development 5
Bertaç Şakir Şahin

Fintech & the Sustainable Development Goals: Financial Inclusion, Development, & Sustainability 11
Chané de Bruyn

Regulatory Frameworks for Fintech in the EU, Germany, and Türkiye 19
Aden Cennet Kılıç

The Role of New Fintech Solutions and Related Regulations in Improving Access to Finance for Consumers and SMEs 43
Iclal Nihal Barac Evci

Business Responsibilities in Achieving Sustainable Development Goals Through Fintech 67
Özcan Serhat Cengiz

State Obligations Under International Law, Role of Fintech in Achieving the SDGs, and Regulations Thereof 95
Hilal Koçali

Conclusion 123
Barış C. Cantürk

Editors and Contributors

About the Editors

Dr. Bertaç Şakir Şahin has been employed as an Associate Professor at the Business Administration Department of Yildiz Technical University since 2018. His M.A. (2019) and Ph.D. (2023) degrees in Accounting and Finance from Marmara University. Sustainability, microfinance, and financial reporting are among his research areas. Besides, he has some studies about CDP disclosures, financial soundness, and sustainable finance. He has published in journals such as the International Journal of Sustainable Economy, Kybernetes, and Management of Environmental Quality. He is actively engaged in several international academic communities. He serves as a deputy head of the "Future Finance Law Hub".

Dr.iur. Barış C. Cantürk is an Assistant Professor of IT Law at Marmara University Faculty of Law, and the founder & director of the Future Finance Law Hub. His academic focus lies on Future Finance Law, which encompasses the intersection of IT Law, Company Law, and Digitalization and Regulation. He has over nine years of professional experience as a research and teaching assistant, attorney at law, and legal counsel. In addition, he has authored several pioneering international publications, including a book, peer-reviewed articles, and symposium proceedings.

He earned his title of Doctor of Law (doctor iuris/dr.iur) at Leipzig University Faculty of Law with the distinction magna cum laude for his dissertation entitled "The Future of Company Law: Decentralized Autonomous Organization—The Existing Legal Wrappers and a New Form Recommendation", supervised by Prof. Dr. Boris P. Paal, M.Jur. (Oxford).

Throughout his academic journey, he has been honoured with multiple awards, scholarships, and fellowships from prestigious institutions such as the Prime Ministry of Türkiye, the Ministry of National Education of Türkiye, the Scientific and Technological Research Council of Türkiye, the German Academic Exchange Service, and the Friedrich Schiedel Foundation.

Contributors

Barış C. Cantürk Assistant Professor of Information and Technology Law, Faculty of Law, Marmara University/Founder and Director, Future Finance Law Hub.

Özcan Serhat Cengiz Research Assistant, Goethe University Leibniz Institute for Financial Research SAFE.

Chané de Bruyn Research Fellow, Centre for Local Economic Development, University of Johannesburg/Member, Future Finance Law Hub.

Iclal Nihal Barac Evci Research Assistant, Department of Commercial Law, Piri Reis University Faculty of Law, Istanbul, Türkiye/Member, Future Finance Law Hub.

Aden Cennet Kılıç Attorney at Law, Istanbul Bar Association/LL.M., TU Dresden/ Member, Future Finance Law Hub.

Hilal Koçali Legal Counsel/Member, Future Finance Law Hub.

Bertaç Şakir Şahin Associate Professor, Department of Business Administration, Yıldız Technical Uni-versity, İstanbul, Türkiye/Deputy Head, Future Finance Law Hub.

Introduction

Bertaç Şakir Şahin

The fusion and advancement of cutting-edge technologies with finance have accelerated the growth of financial technology ("FinTech"). Nevertheless, as a technology-driven financial innovation, FinTech stands apart from conventional financial innovations. FinTech may be characterized as a more profound form of innovation or the most advanced technological development within the financial sector, encompassing cryptocurrencies and blockchain, novel digital advisory and trading platforms, artificial intelligence and machine learning, peer-to-peer ("P2P") lending, equity crowdfunding, and mobile payment solutions (Choudhary & Thenmozhi, 2024). Furthermore, FinTech is inherently disruptive due to its significant innovations in the financial system and associated infrastructure, which, in turn, impact various facets of the economy, society, and sustainability (Offiong et al., 2025).

The intersection of financial technology and sustainable development has become a powerful catalyst in worldwide financial systems. As digital platforms more frequently facilitate access to credit, payments, insurance, and investments, the opportunity to harmonize Fintech advancements with sustainability goals has become increasingly important. This alignment is especially vital given the UN Sustainable Development Goals, which emphasize inclusive, fair, and environmentally responsible economic growth (UNESCAP, 2025).

Nonetheless, the rapid pace of Fintech advancements poses significant regulatory challenges. Policymakers need to find a careful balance between fostering innovation and maintaining financial stability, data protection, and environmental accountability. Regulatory structures are crucial in determining how FinTech aligns with or deviates from the principles of sustainable development.

For that reason, this book examines regulatory approaches to FinTech and sustainability across the European Union ("EU"), Germany, and selected emerging markets, by starting with the introduction of the financial perspective. These jurisdictions

B. Ş. Şahin (✉)
Department of Business Administration, Yıldız Technical University, İstanbul, Turkey
e-mail: bertacsa@yildiz.edu.tr

B. Ş. Şahin and B. C. Cantürk (eds.), *Fintech and Sustainable Development*, Sustainable Finance, https://doi.org/10.1007/978-3-032-31020-0_1

offer distinctive perspectives. While the EU provides a supra-national legal framework with increasing emphasis on Environmental, Social, and Governance ("ESG") integration, Germany leads with its advanced FinTech ecosystem and green finance policies. On the other hand, emerging markets present a dynamic environment where FinTech often addresses infrastructure gaps and financial exclusion, but regulatory capacity may vary. This study examines how regulatory regimes in these jurisdictions influence the development of FinTech, with a view to identifying ways that support or hinder sustainable outcomes. The institutional frameworks, best practices, and regulatory innovations shaping this intersection will be explored through a policy-oriented lens. Hence, the study is structured as follows: (i) a conceptual overview, (ii) region-specific analyses, and (iii) policy recommendations for harmonizing innovation with sustainability goals.

The first chapter establishes the conceptual foundations for connecting financial technology to the Sustainable Development Goals ("SDGs"). It begins by defining the key dimensions of financial inclusion, economic development, and sustainability, outlining how these areas intersect within digital financial ecosystems. The analysis highlights the evolving role of FinTech in advancing the SDGs, including reducing inequalities, promoting gender equality, and expanding access to financial services. Drawing on global data and market trends, the section provides an overview of the FinTech landscape, mapping its growth and potential across various economic contexts. It also presents a critical review of the existing literature to assess how regulatory frameworks shape FinTech's contribution to sustainable outcomes, providing a baseline for the comparative analysis in subsequent sections.

The second chapter examines the legal and institutional frameworks that regulate FinTech innovation in the EU, Germany, and selected emerging economies. It starts with a comprehensive analysis of the current regulatory regime in the EU, including the proposed Payment Services Directive 3 ("PSD3") and the Payment Services Regulation ("PSR"). It evaluates their effects on market integration and innovation. The regulatory context in Germany is reviewed as a national example within the EU system, underscoring its contribution to developing a strong FinTech ecosystem.

The last section analyzes FinTech regulation through the lens of sustainable development, with particular emphasis on financial access, competition, and inclusive growth. It explores how new digital financial services, such as e-money and digital-only banking, are transforming access to finance for underserved segments, particularly small-medium enterprises ("SMEs") and consumers. The role of competition authorities is assessed, focusing on regulatory approaches to market entry, innovation incentives, and antitrust considerations, especially in relation to SDG alignment. A comparative overview of practices in the EU, Türkiye, and emerging markets is provided. This section also evaluates the implementation of regulatory sandboxes and open banking frameworks as instruments for inclusive innovation. Finally, it considers governments' responsibility to leverage FinTech for poverty alleviation, equitable growth, and effective public–private collaboration to achieve the SDGs.

References

Choudhary, P., & Thenmozhi, M. (2024). Fintech and financial sector: ADO analysis and future research agenda. *International Review of Financial Analysis, 93*, 103201.

Offiong, U. P., Szopik-Depczyńska, K., & Ioppolo, G. (2025). FinTech innovations for sustainable development: A comprehensive literature review and future directions. *Sustainable Development, 33*, 1459–1471.

UNESCAP. (2025). *Green growth.* https://www.unescap.org/our-work/environment-development/green-growth. 12 February 2025.

Fintech Investment Trends & Existing Studies on Fintech Regulations and Sustainable Development

Bertaç Şakir Şahin

1 Fintech Investment Trends

As financial technologies continue to advance globally, investments in fintech have grown significantly in recent years, which is crucial for enhancing financial inclusion. However, Global fintech investment in 2024 experienced a notable decline, with total deal value falling to $95.6 billion, the lowest level since 2017, primarily due to macroeconomic uncertainty, geopolitical tensions, and elevated interest rates. Mergers and acquisitions (M&A) dropped from $60.2 billion in 2023 to $49.6 billion in 2024, while venture capital (VC) and private equity (PE) investments also declined to $43.4 billion and $2.6 billion, respectively. The first half of 2024 (H1′24) saw greater investment activity than the second half (H2′24), though a significant uptick was recorded in Q4′24. M&A deal value, for instance, nearly doubled from $7.4 billion in Q3′24 to $14.2 billion in Q4′24, suggesting a moderate recovery in investor sentiment toward the end of the year (KPMG, 2025) (Fig. 1).

This late-year rebound is attributed to the easing of political uncertainty following primary global elections and the gradual reduction in interest rates, which improved capital conditions. Despite overall annual contraction, the Q4′24 resurgence in M&A and VC investment activity indicates a potential shift in market confidence heading into 2025. The observed pattern reflects investor recalibration, with a preference for defensive and strategic investments amid ongoing volatility. These dynamics underscore the cyclical nature of fintech investment flows and the importance of macro-financial stability in sustaining long-term sectoral growth (KPMG, 2025).

The evolution of global fintech investment across six major segments in 2024 reflects a dual trend of defensive consolidation and targeted growth, offering valuable insights into sectoral resilience and strategic repositioning. Notably, the payments segment dominated the landscape, attracting $31 billion in funding, an impressive

B. Ş. Şahin (✉)
Department of Business Administration, Yıldız Technical University, İstanbul, Turkey
e-mail: bertacsa@yildiz.edu.tr

B. Ş. Şahin and B. C. Cantürk (eds.), *Fintech and Sustainable Development*, Sustainable Finance, https://doi.org/10.1007/978-3-032-31020-0_2

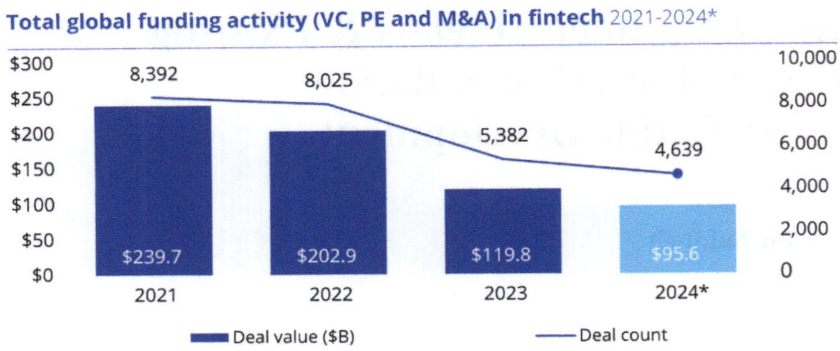

Fig. 1 Total global investment activity in the fintech sector. *Source* KPMG (2025)

rise from $17.2 billion in 2023, mainly driven by megadeals such as the $12.5 billion Worldpay acquisition in H1 and the $6.3 billion take-private of Nuvei in H2, alongside other substantial transactions like Transact Campus and Mynt. In contrast, insurtech experienced a precipitous decline, falling to under $3.1 billion—its lowest level in over a decade—due to the absence of significant M&A activity. However, some mid-sized deals, such as Alan and Zing Health, provided a partial offset. Regtech demonstrated robust growth, with investment increasing from $4.4 billion in 2023 to $7.4 billion in 2024, a surge fueled primarily by Leonard Green's $4 billion acquisition of IRIS Software in H1, even as H2 segment activity stabilized at a more moderate pace (KPMG, 2025).

Within the remaining segments, cybersecurity experienced a contraction, with total investment decreasing from $1.4 billion in 2023 to approximately $890 million in 2024. This decline reflects a strategic shift toward integrative acquisitions by major technology firms and fewer independent large-scale deals, with the largest H2 raise being a $90 million round by Cassava Technologies. Conversely, wealthtech exhibited a strong recovery, doubling investment from a more than decade-low $190 million in 2023 to $400 million in 2024, primarily in H2, supported by significant funding rounds led by Edward Jones ($73.9 million), Bitwise ($60 million), Baseload Capital ($58.5 million), and Neo Asset Management ($47.9 million). Finally, the digital assets and currencies sector reached $9.1 billion in global investment, up from $8.7 billion in 2023, with the U.S. market driving growth through megadeals like Stripe's $1.1 billion acquisition of Bridge, a $525 million raise by Praxis, and other notable regional investments. Collectively, these dynamics portray a fintech ecosystem that is simultaneously retrenching in certain areas while selectively expanding in others, indicative of a recalibrated investor strategy amid ongoing macroeconomic uncertainty (KPMG, 2025).

2 Existing Studies on Fintech Regulations and Sustainable Development

Financial markets can play an important role in transforming savings into investment and in efficient resource allocation. Market imperfections and asymmetric information can hinder the efficient allocation of capital, leaving some firms and households excluded from formal financial systems and thereby impeding inclusive economic growth. Globally, the persistent challenges of financial exclusion and income inequality have intensified concerns over the finance–development nexus. The UN 2030 Agenda for Sustainable Development highlights the pivotal role of financial inclusion in achieving the Sustainable Development Goals, particularly in reducing inequality (SDG 10) (Klapper et al., 2016).

Fintech has the potential to significantly improve financial inclusion by using technology to eliminate traditional obstacles to accessing finance. By leveraging digital platforms, mobile banking, and algorithm-driven credit evaluation tools, fintech enables marginalized groups, especially those in remote or economically disadvantaged areas, to access financial services without physical bank branches. Fintech solutions may lower transaction fees, streamline the onboarding process, and provide more adaptable, customer-focused products designed to meet a variety of financial requirements (Philippon, 2019). The literature provides empirical evidence that fintech applications will have an important role in combating inequality (Abor et al., 2018; Asongu & Odhiambo, 2018). Additionally, there is a growing body of literature on the role of financial technologies in expanding financial access for disadvantaged groups.

Studies examine whether financial inclusion can be further increased through fintech's digital channels. Fintech is anticipated to address deficiencies in both payment processing and lending, particularly in areas where conventional financial service delivery is lacking. Fuster et al. (2019) demonstrate that fintech frequently acts as a supplement to, rather than a replacement for, traditional banking services. In contrast, Hau et al. (2024) and Frost et al. (2019) contend that fintech and large technology lenders cater to borrowers whom banks have historically overlooked.

The literature is limited in its empirical evidence on the relationship between fintech and financial inclusion.

Demir et al. (2022) examine the interrelationships among FinTech, financial inclusion, and income inequality across a panel of 140 countries, using data from the Global Findex surveys conducted in 2011, 2014, and 2017. The study posits that FinTech influences income inequality both directly and indirectly via its impact on financial inclusion. Using quantile regression, the authors assess whether these effects vary across countries with differing levels of income inequality. The findings provide novel evidence that financial inclusion serves as a critical transmission channel through which FinTech helps reduce income inequality. Moreover, while financial inclusion significantly reduces inequality across all quantiles of the distribution, its impact is more pronounced in higher-income countries.

Tok and Heng (2022) investigate the role of fintech in promoting financial inclusion by utilizing Global Findex data alongside emerging fintech indicators. Their findings reveal that fintech adoption shows a stronger positive correlation with digital financial inclusion than with traditional measures of financial inclusion. In the second phase of their analysis, the authors explore the relationship between fintech usage and the three dimensions of the digital divide: gender, income (class), and rural–urban. The results suggest that increased fintech usage is significantly associated with reductions in the income and rural divides, whereas no significant effect is observed on the gender divide.

Kanga et al. (2022) find that ATMs, related digital networks, mobile phones, and payment systems increase per capita income through financial inclusion in 137 countries. Rosyadah et al. (2021) highlight the importance of Fintechs for the financial inclusion of Micro, Small, and Medium Enterprises using a survey methodology. Esmaeilpour Moghadam and Karami et al. (2023) analyze the impact of Fintech applications on women's financial inclusion in 113 countries. The findings show that fintech positively impacts women's access to finance. Lu (2024) emphasize that fintech can contribute to green finance through financial inclusion.

Ololade (2024) analyzes the contribution of fintech to financial inclusion in the U.S. and African countries. The paper identifies regulatory hurdles, digital literacy, and infrastructure limitations, providing a nuanced understanding of the Fintech ecosystem. Ediagbonya and Tioluwani (2023) examine the impact of fintech on inclusion in emerging countries with content analysis. Empirical findings explain the impact of fintech on financial inclusion through financial literacy, information asymmetry, and infrastructure adequacy. Goswami et al. (2022) show the importance of social and behavioral factors in the impact of fintech on financial inclusion in India. Azmeh (2025) analyze data from 108 low to upper-middle-income countries spanning 2011, 2014, 2017, and 2021. The study's findings highlight fintech's transformative impact on financial inclusion, with mobile banking complementing traditional banking to reduce poverty rates.

The literature is limited in its analysis of the relationship between fintech and financial inclusion, focusing primarily on supply and demand. Ashenafi and Yan (2023) analyze the impact of fintech on financial inclusion across 49 countries, focusing on supply and demand. According to the survey results, the quality of institutions contributes to the impact of fintech on financial inclusion. Loo (2019) examines the relationship between fintech and financial inclusion in ASEAN countries with supply, demand, and risk dimensions. The demand variable is financial accounts, while the supply variables are infrastructure and the Fintech ecosystem. Institutional factors are analyzed as risk variables. Velazquez et al. (2022) analyze the impact of fintech on financial inclusion in Peru using six demand-side indicators, selected for each country from the Global Findex Database. The findings indicate that Fintech services do not significantly impact financial inclusion.

Finally, it should be emphasized that fintech regulations can play a critical role in aligning financial innovation with sustainable development goals. By ensuring transparency, consumer protection, and systemic stability, regulatory frameworks can foster trust and long-term participation in digital financial systems. Well-designed regulations also encourage the responsible deployment of technologies that promote financial inclusion, green finance, and equitable access to capital, key enablers of the United Nations Sustainable Development Goals (SDGs). Conversely, overly restrictive or fragmented regulations may hinder innovation and limit fintech's potential to contribute to inclusive and sustainable economic growth (Arner et al., 2020; Treleaven, 2015). Despite this, the literature handling the financial inclusion relationship of fintech regulations is limited (Arner et al., 2020; Lu, 2024; Widjaja, 2025). Studies focus on the fintech and banking performance of fintech regulations rather than the fintech regulation-financial inclusion relationship (Goo & Heo, 2020; Khan et al., 2023; Nguyen et al., 2022).

References

Abor, J. Y., Amidu, M., & Issahaku, H. (2018). Mobile telephony, financial inclusion and inclusive growth. *Journal of African Business, 19*(3), 430–453.

Arner, D. W., Buckley, R. P., Zetzsche, D. A., & Veidt, R. (2020). Sustainability, FinTech and financial inclusion. *European Business Organization Law Review, 21*, 7–35.

Ashenafi, B. B., & Yan, D. (2023). Financial intermediation, inclusion, fintech, and income inequality in Africa: Robust evidence from the supply and demand side data. *Economic Notes, 52*(2), e12221.

Asongu, S. A., & Odhiambo, N. M. (2018). Human development thresholds for inclusive mobile banking in developing countries. *African Journal of Science, Technology, Innovation and Development, 10*(6), 735–744.

Azmeh, C. (2025). Bridging divides: The role of fintech and financial inclusion in reducing poverty and inequality in developing countries. *Innovation and Development*, 1–20. https://doi.org/10.1080/2157930X.2025.2467515

Choudhary, P., & Thenmozhi, M. (2024). Fintech and financial sector: ADO analysis and future research agenda. *International Review of Financial Analysis, 93*, 103201.

Demir, A., Pesqué-Cela, V., Altunbas, Y., & Murinde, V. (2022). Fintech, financial inclusion and income inequality: A quantile regression approach. *The European Journal of Finance, 28*(1), 86–107.

Ediagbonya, V., & Tioluwani, C. (2023). The role of fintech in driving financial inclusion in developing and emerging markets: Issues, challenges and prospects. *Technological Sustainability, 2*(1), 100–119.

Esmaeilpour Moghadam, H., & Karami, A. (2023). Financial inclusion through fintech and women's financial empowerment. *International Journal of Social Economics, 50*(8), 1038–1059.

Frost, J., Gambacorta, L., Huang, Y., Shin, H. S., & Zbinden, P. (2019). BigTech and the changing structure of financial intermediation. *Economic Policy, 34*(100), 761–799.

Fuster, A., Plosser, M., Schnabl, P., & Vickery, J. (2019). The role of technology in mortgage lending. *The Review of Financial Studies, 32*(5), 1854–1899.

Goo, J. J., & Heo, J. Y. (2020). The impact of the regulatory sandbox on the fintech industry, with a discussion on the relation between regulatory sandboxes and open innovation. *Journal of Open Innovation: Technology, Market, and Complexity, 6*(2), 43.

Goswami, S., Sharma, R. B., & Chouhan, V. (2022). Impact of financial technology (Fintech) on financial inclusion (FI) in Rural India. *Universal Journal of Accounting and Finance, 10*(2), 483–497.

Hau, H., Huang, Y., Lin, C., Shan, H., Sheng, Z., & Wei, L. (2024). FinTech credit and entrepreneurial growth. *The Journal of Finance, 79*(5), 3309–3359.

Kanga, D., Oughton, C., Harris, L., & Murinde, V. (2022). The diffusion of fintech, financial inclusion and income per capita. *The European Journal of Finance, 28*(1), 108–136.

Khan, H. H., Khan, S., & Ghafoor, A. (2023). Fintech adoption, the regulatory environment and bank stability: An empirical investigation from GCC economies. *Borsa Istanbul Review, 23*(6), 1263–1281.

Klapper, L., El-Zoghbi, M., & Hess, J. (2016). *Achieving the sustainable development goals: The role of financial inclusion.* CGAP.

KPMG. (2025). *Pulse of Fintech. H2'24.* https://assets.kpmg.com/content/dam/kpmgsites/xx/pdf/2025/02/pulse-of-fintech-h2-2024.pdf.coredownload.pdf. 12 May 2025.

Loo, M. K. L. (2019). Enhancing financial inclusion in ASEAN: Identifying the best growth markets for fintech. *Journal of Risk and Financial Management, 12*(4), 181.

Lu, L. (2024). The law of fintech: How Artificial intelligence and innovative technologies contribute to a sustainable financial industry and its effective regulation. In *Artificial intelligence, finance, and sustainability: Economic, ecological, and ethical implications* (pp. 243–263). Springer Nature Switzerland.

Nguyen, L., Tran, S., & Ho, T. (2022). Fintech credit, bank regulations and bank performance: A cross-country analysis. *Asia-Pacific Journal of Business Administration, 14*(4), 445–466.

Offiong, U. P., Szopik-Depczyńska, K., & Ioppolo, G. (2025). FinTech innovations for sustainable development: A comprehensive literature review and future directions. *Sustainable Development, 33,* 1459–1471.

Ololade, Y. J. (2024). Conceptualizing Fintech innovations and financial inclusion: Comparative analysis of African and U.S. initiatives. *Finance & Accounting Research Journal, 6*(4), 546–555.

Philippon, T. (2019). *On fintech and financial inclusion* (No. w26330). National Bureau of Economic Research.

Rosyadah, K., Budiandriani, B., & Hasrat, T. (2021). The role of fintech: Financial inclusion in MSME's: Case study in Makassar City. *Jurnal Manajemen Bisnis, 8*(2), 268–275.

Tok, Y. W., & Heng, D. (2022). *Fintech: Financial inclusion or exclusion?* International Monetary Fund. https://www.imf.org/en/Publications/WP/Issues/2022/05/06/Fintech-Financial-Inclusion-or-Exclusion-517619. 12 February 2025.

Treleaven, P. (2015). Financial regulation of FinTech. *Journal of Financial Perspectives.* https://papers.ssrn.com/sol3/papers.cfm?abstract_id=3084015. 5 March 2025.

Velazquez, P. V., Bobek, V., Vide, R. K., & Horvat, T. (2022). Lessons from remarkable Fintech companies for the financial inclusion in Peru. *Journal of Risk and Financial Management, 15*(2), 62.

Widjaja, G. (2025). Analyzing the impact of sandbox regulation on fintech innovation and financial inclusion. *International Journal of Society Reviews, 3*(5), 860–866.

Fintech & the Sustainable Development Goals: Financial Inclusion, Development, & Sustainability

Chané de Bruyn

1 Introduction to Fintech and the Sustainable Development Goals

Fintech significantly contributes to advancing the SDGs by enhancing financial inclusion, promoting economic development, and supporting sustainability efforts. Fintech, which stands for financial technology, involves the application of innovative technologies to deliver financial services, frequently challenging traditional financial frameworks (Lee, 2017; Yap, 2023). It covers a broad array of services, including mobile payments, online banking, insurance, wealth management, cryptocurrencies, and international payments (Hsieh & Li, 2017). Typically, fintech firms leverage technological advancements such as blockchain, artificial intelligence, and data analytics to develop financial products and services that are more efficient and accessible (Hua et al., 2019; Mohamed, 2021).

Fintech innovations have significantly improved access to financial services, particularly in developing regions (Azmeh & Al-Raeei, 2024). These advancements include digital payment systems, peer-to-peer lending, and mobile banking, all of which help bridge the gap for marginalised populations (Kumar et al., 2024; Sharma et al., 2025). Abubakar and Handayani (2018) write that by enhancing the efficiency and reach of financial services, fintech also plays a crucial role in driving economic development. By empowering people to become more economically active, fintech allows people greater opportunities to engage in the economy and enhance their quality of life (Raffaelli et al., 2025). Additionally, it can support gender equality by granting women access to essential financial services, which play a vital role in strengthening their economic independence (Chitimira & Warikandwa, 2023).

C. de Bruyn (✉)
Centre for Local Economic Development, University of Johannesburg, Johannesburg, South Africa
e-mail: chanedb@uj.ac.za

B. Ş. Şahin and B. C. Cantürk (eds.), *Fintech and Sustainable Development*, Sustainable Finance, https://doi.org/10.1007/978-3-032-31020-0_3

All of these factors make fintech particularly important in realising the road towards achieving the SDGs, especially in the current digital age. As mentioned, fintech has a key role in advancing financial inclusion by providing access to services that traditional banks often overlook, particularly in underserved regions (Nair & Lopez, 2024). This is especially significant for advancing the SDGs related to poverty alleviation and economic growth (Kesavan & Polisetty, 2025). For example, in Saudi Arabia, various fintech initiatives are closely linked to SDGs such as No Poverty (SDG 1), Zero Hunger (SDG 2), and Decent Work and Economic Growth (SDG 8), as they enhance access to critical services while driving economic development (Noreen, 2024). Similarly, Raffaelli et al. (2025) explain how across Latin America, fintech has been shown to promote financial inclusion and support several SDGs, including Industry, Innovation, and Infrastructure (SDG 9).

Moreover, fintech contributes to economies by boosting productivity, fostering entrepreneurship, and driving technological advancements (Choudhary et al., 2025). It particularly empowers SMEs by providing easier access to financing, which is vital for their growth and sustainability (Junarsin et al., 2023). A study by Ahiase et al. (2024) explains that in ASEAN countries, the adoption of fintech has demonstrated its potential to improve environmental, economic, and social sustainability, especially by helping narrow the income gap. Additionally, in Africa, the relationship between fintech and financial inclusion is highly beneficial for sustainable development, as noted by Tidjani and Madouri (2024); however, this interaction can also have negative impacts. These challenges are elaborated on in the next section.

Additionally, fintech supports the SDGs by promoting green finance and facilitating the transition to clean energy (Tiwari et al., 2024). Wu et al. (2024) add that it directs financial resources towards environmentally friendly projects while also strengthening the enforcement of environmental policies. Fintech has also been shown to aid in the shift of more traditional linear economies to circular economies (Kesavan & Polisetty, 2025; Phukan & Kumar, 2024). Xu et al. (2025) argue that fintech accelerates green finance initiatives by enhancing financial efficiency and catalyzing eco-friendly innovation, with greater impacts in advanced financial markets. Furthermore, by promoting digital transactions, fintech reduces reliance on paper-based systems, thereby contributing to reduced deforestation and lower carbon footprints associated with paper production (Zhe et al., 2024). The incorporation of fintech into financial frameworks, coupled with its role in advancing green finance and sustainable solutions, serves as a cornerstone in fostering environmentally conscious economies and attaining global sustainability objectives. The subsequent section delves deeper into fintech's role in promoting sustainable development.

2 Understanding the Relationship Between Fintech and the SDGs

As the previous section indicated, fintech plays a crucial role in advancing the SDGs by offering innovative solutions that could increase financial inclusion, promote economic growth, and support environmental sustainability. This being said, the connection between fintech and the SDGs is complex, influencing various facets of sustainable development.

Innovations in fintech, such as mobile money, digital banking, and blockchain technology, have greatly enhanced access to financial services for underserved populations, especially in developing nations (Carè et al., 2024). This increased accessibility is vital for fostering economic growth and improving livelihoods, directly supporting SDG 1 (No Poverty) and SDG 8 (Decent Work and Economic Growth) (Kesavan & Polisetty, 2025; Noreen, 2024). By providing financial services to micro, small, and medium-sized enterprises (MSMEs), fintech encourages entrepreneurship and overall economic development, thereby contributing to achieving SDG 8 (Ferrata, 2019). Additionally, Carè et al. (2024) suggest that fintech has the potential to bridge gender gaps in financial inclusion, thereby promoting SDG 5 (Gender Equality). Moreover, by offering financial services to marginalized communities, fintech helps reduce inequalities within and between countries, aligning with SDG 10 (Reduced Inequalities) (Kuzmina et al., 2024; Noreen, 2024).

Furthermore, as mentioned in the previous section, fintech promotes green finance by facilitating environmentally friendly investments and advancing low-carbon technologies, advancing SDG 13 (Climate Action) and the broader objective of sustainable development (Omeragic et al., 2024; Saqib et al., 2023). Saqib et al. (2023) state that there is a clear correlation between fintech and renewable energy, underscoring fintech's role in driving a green transition and further supporting SDGs focused on environmental sustainability. Also, as is most often the case, fintech initiatives rely on collaborations among governments, financial institutions, and tech firms, which are essential for achieving SDG 17 (Partnerships for the Goals) (Noreen, 2024). Further examples of how fintech solutions are directly contributing to achieving various SDGs by leveraging technology to address financial, environmental, and social challenges are presented in Table 1.

Despite fintech offers significant opportunities to advance sustainable development, it also poses a range of challenges and risks that must be carefully managed. To fully harness the benefits of fintech, developing countries require significant infrastructure investments (Hinson et al., 2019). This includes establishing digital identity systems, interoperable electronic payment solutions, and robust digital financial markets (Arner et al., 2020). Studies such as Razletovskaia (2020) argue that the effective implementation of fintech requires coordinated efforts across sectors, including digital technology, finance, telecommunications, IT, and regulatory bodies. This cross-sector collaboration is essential for achieving the SDGs (Razletovskaia, 2020). However, coordinating such efforts can be particularly difficult in regions with limited government capacity. Insufficient technical knowledge, coupled with a

Table 1 Examples of fintech solutions and their link to SDGs

Fintech solution	Supported SDGs	Description
Mobile Banking and Digital Wallets	SDG 1, SDG 8	Improves financial inclusion for marginalised populations
Microfinance Services	SDG 1, SDG 8	Allows for smaller loans to individuals and small businesses
Green Bonds and Green Insurance	SDG 7, SDG 13	Finances sustainable energy projects
IoT and AI-Enabled Green Fintech	SDG 7, SDG 13	Improves efficiency and the impact of sustainable investments
Blockchain for Supply Chains	SDG 12	Ensures transparency and efficiency in supply chains
Robo-Advisors for Environmentally Friendly Investments	SDG 13	Allocates funds to sustainable projects
Crowdfunding and Alternative Financing	SDG 8, SDG 9	Supports entrepreneurship and innovation
Digital Identity and e-KYC Systems	SDG 8	Enhances financial inclusion and access to finance
ESG Reporting and Corporate Governance	SDG 12, SDG 16	Improves sustainability practices and transparency

Source Arner et al. (2020), Jha et al. (2022), Federico and Adamo (2024), Misra et al. (2024) and Omeragic et al. (2024)

shortage of skilled professionals, represents a significant obstacle to the widespread adoption of fintech solutions (Amri et al., 2025). Therefore, there is a pressing need for capacity building to enhance human resources and technical expertise (Amri et al., 2025; Nguyen et al., 2020).

The rapid advancement of fintech often outpaces regulators' ability to adapt, leading to reactive and sometimes insufficient responses (Madir, 2021; Vijayagopal et al., 2024). Lim and Low (2019) argue that traditional regulatory measures are often viewed as "too little, too late," especially for new technologies such as cryptocurrencies and digital assets. Moreover, there is a noticeable disparity in the pace of fintech adoption and regulation across various countries. Developed nations like the U.S. and the UK are at the forefront of innovation, while many developing countries grapple with challenges in digital infrastructure and financial literacy (Vijayagopal et al., 2024). To address the swift evolution of fintech, some jurisdictions have adopted regulatory sandboxes. These allow fintech developers to test their products under regulatory oversight (Lim & Low, 2019). This collaborative approach is designed to ensure compliance while also promoting innovation (Lim & Low, 2019).

There are also significant ethical challenges associated with fintech's role in advancing the SDGs. One primary concern highlighted by Pawlak (2024) is the risk of greenwashing, where companies falsely portray their environmental initiatives. This issue has gained prominence as fintech firms are subject to heightened regulatory scrutiny, particularly under directives such as the EU's CSRD (Pawlak, 2024). Furthermore, the integration of fintech in sectors such as healthcare in Sub-Saharan

Africa raises ethical considerations around privacy, data security, and equitable access to technology (Cambaza, 2023).

One of the significant ethical challenges facing the fintech industry is the lack of standardized ESG metrics (Roy & Vasa, 2025). This inadequacy makes it challenging to assess and compare sustainability efforts across various fintech solutions. The absence of uniform reporting can lead to inconsistent information and potentially misleading sustainability claims (Roy & Vasa, 2025). Although fintech has the potential to improve financial inclusion, it also raises ethical questions about equitable access. For example, Chand et al. (2025) note that in Pakistan's informal sector, small businesses often struggle to access formal financial services despite advancements in fintech. This situation underscores the urgent need for inclusive fintech solutions that truly serve underserved communities without deepening existing inequalities (Chand et al., 2025).

Moreover, integrating AI into fintech raises significant ethical concerns about data privacy and security (Khang et al., 2025). It is crucial to adopt responsible AI practices and safeguard user data to ensure that fintech solutions contribute positively to sustainable development (Khang et al., 2025). In addition, as fintech solutions frequently require substantial data collection, this raises significant concerns about the use, storage, and sharing of such data (Aldboush & Ferdous, 2023; Nurmara et al., 2023). The rapid proliferation of fintech applications has increased security vulnerabilities, rendering them more susceptible to cyberattacks and data breaches (Namchoochai et al., 2020; Shafik, 2025). Therefore, ethical considerations must include assurances that these technologies do not unintentionally exclude or exploit vulnerable populations and that user privacy is protected.

3 Conclusion

In conclusion, the relationship between fintech and the SDGs is transformative and multifaceted. Literature shows that fintech plays a vital role in promoting financial inclusion, economic participation, and environmentally responsible innovation across various regions. Through digital payments, mobile banking, alternative financing, and green financial instruments, fintech enhances access to essential financial services, supports entrepreneurship, and directs investments toward low-carbon and socially beneficial activities. These contributions advance several SDGs, particularly those focused on poverty reduction, gender equality, economic growth, innovation, and climate action.

However, studies reveal that the potential of fintech is not automatic or evenly distributed. To fully realize its benefits, substantial investments are needed in digital infrastructure, human capital, and regulatory capacity areas where many developing countries still encounter significant challenges. The rapid pace of technological change also poses challenges to traditional regulatory frameworks, resulting in gaps in oversight and uneven adoption of fintech across different countries. Ethical concerns such as data privacy, cybersecurity risks, greenwashing, and unequal access to digital services further complicate fintech's role in sustainable development. Without strong governance mechanisms and coordinated collaboration across sectors, these risks could undermine the very SDGs that fintech strives to support.

In summary, fintech represents both an opportunity and a responsibility, as it is a powerful tool that can accelerate progress toward the SDGs, but it also requires intentional policy design, institutional coordination, and ethical safeguards to ensure inclusive and sustainable outcomes. These complexities set the scene for the next section, which will explore in greater detail how fintech contributes to sustainable development and the conditions needed to maximize its positive impact.

References

Abubakar, L., & Handayani, T. (2018). Financial technology: Legal challenges for Indonesia financial sector. *IOP Conference Series: Earth and Environmental Science, 175*(1), 012204.

Ahiase, G., Andriana, D., Widyaningsih, A., Heryana, T. & Purnomo, B. S. (2024). Financial technology and sustainable development in ASEAN region: Role of income inequality. *Jurnal Ekonomi Malaysia, 58.*

Aldboush, H. H., & Ferdous, M. (2023). Building trust in fintech: An analysis of ethical and privacy considerations in the intersection of big data, AI, and customer trust. *International Journal of Financial Studies, 11*(3), 90.

Amri, M. C. E., Mohammed, M. O., & AlKhalili, M. M. (2025). Fintech adoption and its investment impact in Islamic social finance: The case of Zakat. *Qudus International Journal of Islamic Studies, 12*, 213–254.

Arner, D. W., Buckley, R. P., Zetzsche, D. A., & Veidt, R. (2020). Sustainability, FinTech and financial inclusion. *European Business Organization Law Review, 21*, 7–35.

Azmeh, C., & Al-Raeei, M. (2024). Exploring the dual relationship between fintech and financial inclusion in developing countries and their impact on economic growth: Supplement or substitute? *PLoS ONE, 19*(12), e0315174.

Cambaza, E. (2023). The role of FinTech in sustainable healthcare development in sub-Saharan Africa: A narrative review. *FinTech, 2*(3), 444–460.

Carè, R., Boitan, I. A., Stoian, A. M., & Fatima, R. (2024). Exploring the landscape of financial inclusion through the lens of financial technologies: A review. *Finance Research Letters, 72*, 106500.

Chand, A., Liu, D., Zulfiqar, M., Ullah, M. R., & Khan, M. J. (2025). Perceived quality in Fintech services: Expanding UTAUT2 and the Delone and McLean Information System Success Models. *Business Process Management Journal.*

Chitimira, H., & Warikandwa, T. V. (2023). Financial inclusion as an enabler of United Nations sustainable development goals in the twenty-first century: An introduction. *Financial inclusion and digital transformation regulatory practices in selected SADC countries: South Africa, Namibia, Botswana and Zimbabwe* (pp. 1–22). Springer International Publishing.

Choudhary, P., Ghosh, C., & Thenmozhi, M. (2025). Impact of fintech and financial inclusion on sustainable development goals: Evidence from cross country analysis. *Finance Research Letters, 72*, 106573.

Federico, D., & Adamo, R. (2024). Alternative finance. A framework for innovative and sustainable business models. In *Routledge international studies in money and banking* (pp. 134–150).

Ferrata, L. (2019). Digital financial inclusion–an engine for "leaving no one behind." *Public Sector Economics, 43*(4), 445–458.

Hinson, R., Lensink, R., & Mueller, A. (2019). Transforming agribusiness in developing countries: SDGs and the role of FinTech. *Current Opinion in Environmental Sustainability, 41*, 1–9.

Hsieh, K. H., & Li, E. Y. (2017). Progress of Fintech industry from venture capital point of view. In *Proceedings of the International Conference on Electronic Business (ICEB)*, Dubai, UAE.

Hua, X., Huang, Y., & Zheng, Y. (2019). Current practices, new insights, and emerging trends of financial technologies. *Industrial Management & Data Systems, 119*(7), 1401–1410.

Jha, B., Giri, P., Jha, D., & Badhera, U. (2022). Unlocking IoT: AI-enabled Green FinTech innovations. In *AI-enabled Agile Internet of Things for sustainable FinTech ecosystems* (pp. 1–22). IGI Global.

Junarsin, E., Hanafi, M. M., Iman, N., Arief, U., Naufa, A. M., Mahastanti, L., & Kristanto, J. (2023). Can technological innovation spur economic development? The case of Indonesia. *Journal of Science and Technology Policy Management, 14*(1), 25–52.

Kesavan, V., & Polisetty, A. (2025). An extensive examination of the influence of financial technology (Fintech) on advancing financial inclusion: A bibliometric investigation. *Discover Sustainability, 6*(1), 1–37.

Khang, A., Jadhav, B., Hajimahmud, V. A., & Satpathy, I. (2025). *Synergy of AI and fintech in the digital gig economy.* CRC Press.

Kumar, L., Katyayani, J., Venkatesh, D., Sunkara, S., Gowthami, K., & Rani, C. (2024). Exploring the opportunities of fintech services. *Communications on Applied Nonlinear Analysis, 31*(5S), 79–90.

Kuzmina, J., Romānova, I., Natrins, A., Spilbergs, A., & Mavlutova, I. (2024). Innovative digital technologies in finance for promoting sustainable development goals: Baltic economies compared to OECD countries. *WSEAS Transactions on Environment and Development, 20*, 973–986.

Lee, S. (2017). Evaluation of mobile application in user's perspective: Case of P2P lending Apps in FinTech industry. *KSII Transactions on Internet and Information Systems (TIIS), 11*(2), 1105–1117.

Lim, B., & Low, C. (2019). Regulatory sandboxes. In *FinTech: Law and regulation* (pp. 302–325). Edward Elgar Publishing Ltd.

Lu, L. (2024). The law of Fintech: How artificial intelligence and innovative technologies contribute to a sustainable financial industry and its effective regulation. In *Artificial intelligence, finance, and sustainability: Economic, ecological, and ethical implications* (pp. 243–263). Springer Nature Switzerland.

Madir, J. (2021). *FinTech: Law and regulation.* Edward Elgar Publishing.

Misra, N., Gupta, S., & Rao, T. J. (2024). Fintech's role in green finance: Procuring funds for sustainable energy. *International Research Journal of Multidisciplinary Scope, 5*(02), 290–312.

Mohamed, H. (2021). *Beyond fintech: Technology applications for the Islamic economy* (pp. 1–250). World Scientific Publishing Co.

Nair, B. A., & Lopez, K. (2024). Forward and backward linkages of financial inclusion, fintech, and sustainable development goals: A bibliometric analysis and concept mapping. In *Financial inclusion: In Pursuit of UN Sustainable Development Goal 8.10* (pp. 29–43). Springer Nature Switzerland.

Namchoochai, R., Kiattisin, S., Darakorn Na Ayuthaya, S., & Arunthari, S. (2020). Elimination of fintech risks to achieve sustainable quality improvement. *Wireless Personal Communications, 115*(4), 3199–3214.

Nguyen, D. D., Dinh, H. C., & Nguyen, D. V. (2020). Promotion of fintech application for the modernization of banking-finance system in Vietnam. *The Journal of Asian Finance, Economics and Business, 7*(6), 127–131.

Noreen, U. (2024). Mapping of FinTech ecosystem to Sustainable Development Goals (SDGs): Saudi Arabia's landscape. *Sustainability, 16*(21), 9362.

Nurmara, M. A. K., Hakim, M. N., Ardy, O. H. C., Jeffrey, R., Setiono, V. A., Kanigoro, B., & Irwansyah, E. (2023). A review of security in financial technology. *Procedia Computer Science, 227*, 958–965.

Omeragic, N., Zaimovic, A., & Zaimovic, T. (2024). FinTech and climate action, and affordable and clean energy. *Procedia Computer Science, 236*, 273–280.

Pawlak, K. (2024). Greenwashing in the FinTech sector: The role of the EU Corporate Sustainability Reporting Directive (CSRD). In *European conference on innovation and entrepreneurship* (pp. 1023–1031). Academic Conferences International Limited.

Phukan, P. K., & Kumar, P. P. (2024). The intersection of fintech and sustainability: A catalyst for positive change. In *The adoption of fintech* (pp. 98–113). Productivity Press.

Raffaelli, P., Correa-García, J. A., & Verón, C. S. (2025). Financial inclusion and fintech: Catalysts for the sustainable development goals in Latin America. *Retos, 15*(29), 47–63.

Razletovskaia, V. (2020). International coordination and national institutional facilitating mechanisms for financial technology development, for the sustainable development support. In *E3S Web of Conferences* (Vol. 208, p. 03041). EDP Sciences.

Roy, J. K., & Vasa, L. (2025). Financial technology and environmental, social and governance in sustainable finance: A bibliometric and thematic content analysis. *Discover Sustainability, 6*(1), 1–22.

Saqib, N., Mahmood, H., Murshed, M., Duran, I. A., & Douissa, I. B. (2023). Harnessing digital solutions for sustainable development: A quantile-based framework for designing an SDG framework for green transition. *Environmental Science and Pollution Research, 30*(51), 110851–110868.

Shafik, W. (2025). Security, privacy, and trust in fintech. In *FinTech and financial inclusion* (pp. 216–233). Routledge.

Sharma, V., Gupta, M., Arora, N. & Shaikh, A. A. (2025). *FinTech and financial inclusion: Leveraging digital finance for economic empowerment and sustainable growth* (pp. 1–301). Taylor & Francis.

Tidjani, C., & Madouri, A. (2024). Fintech, financial inclusion, and sustainable development in the African region. *Frontiers in Applied Mathematics and Statistics, 10*, 1276218.

Tiwari, S., Cheong, C. W., Mey, L. S., & Saji, T. G. (2024). Does fintech really matter for energy, economy and environment? From the lenses of SDG-7, SDG-8, SDG-13, COP27 and COP28. *Resources Policy, 98*, 105318.

Vijayagopal, P., Jain, B., & Ayinippully Viswanathan, S. (2024). Regulations and fintech: A comparative study of the developed and developing countries. *Journal of Risk & Financial Management, 17*(8).

Wu, B., Ding, Y., Xie, B., & Zhang, Y. (2024). FinTech and inclusive green growth: A causal inference based on double machine learning. *Sustainability, 16*(22).

Xu, J., Li, T., Chen, F., & Quan, Y. (2025). Does Fintech Foster the development of green finance? *Emerging Markets Finance and Trade, 61*(6), 1771–1781.

Yap, S. T. L. (2023). Fintech market development in China. In *Contemporary strategic Chinese American business negotiations and market entry* (pp. 471–493). Springer Nature Singapore.

Zhe, D., Su, N., Zhu, X., Mahmoud, H. A., & Akhtar, T. (2024). Non-linear relationship between FinTech, natural resources, green innovation and environmental sustainability: Evidence from panel smooth transition regression model. *Resources Policy, 91*, 104902.

Regulatory Frameworks for Fintech in the EU, Germany, and Türkiye

Aden Cennet Kılıç

1 Fintech Landscape and Development in Germany

Over the last few years, financial technologies ("fintech") have emerged as a hot topic among market actors, financial institutions, and policymakers in Germany. In this context, regulatory authorities have taken proactive steps to support and monitor the sector. In 2022, the German Federal Financial Supervisory Authority ("BaFin") initiated the Fintech Innovation Hub, where BaFin shares regular updates on developments affecting fintech companies. Through this initiative, BaFin aims to foster dialogue between fintech companies ("fintechs") and the regulator, providing both guidance and transparency (FinTech Innovation Hub, BaFin, n.d.)

This regulatory attention aligns with the rapid expansion of fintech firms in Germany. According to the Germany Trade & Invest report, there are over 1,000 fintech firms active in Germany (Germany Trade & Invest, 2025). The concentration of these firms also highlights regional hubs of innovation: The capital Berlin hosts by far the most fintech companies, followed by Frankfurt, Munich, Hamburg, and Cologne (Fintechhauptstadt Berlin—Eine vergleichende Analyse, 2025).

In addition to the high volume, the spread of the fintech operations also appears to be wide. Latest reports reveals that fintech companies operates mainly in four sectors: Business tools (accounting, controlling & tax/risk & compliance), Personal tools (personal financial management/peer-to-peer), Single service providers (asset management & investment/payments, orders & cash/credit & factoring), Full-service providers (banking & API banking/insurance/DeFi) (FinTech Made in Germany, 2024). This diversity demonstrates the wide-reaching impact of fintechs on both business-to-business (B2B) and business-to-consumer (B2C) markets.

A. C. Kılıç (✉)
Istanbul Bar Association, Istanbul, Turkey
e-mail: adenkilic@gmx.de

TU Dresden, Dresden, Germany

B. Ş. Şahin and B. C. Cantürk (eds.), *Fintech and Sustainable Development*, Sustainable Finance, https://doi.org/10.1007/978-3-032-31020-0_4

Even though the fintech companies possess a strong position in Germany, challenges arose that were detrimental to the fast growth of the fintech businesses. In 2023, over $ 1 billion was invested in German fintechs; however, due to the economic downturn, rising inflation, and investor hesitation, growth declined significantly (Branchencenter Updates, 2025).

These market conditions also influenced investment patterns. While venture capital funding for fintech institutions dropped in 2023 due to market challenges compared to the prior year, funding priorities also changed. Additionally, investors preferred B2B companies over B2C, given their steady profits and larger budgets (Branchencenter Updates, 2025).

Consequently, fintech companies responded by adapting their business strategies. The challenges in the market pushed fintech initiatives to collaborate with traditional banks to provide new services, such as embedded finance (integrating financial services into other products) and banking-as-a-service (allowing other businesses to offer banking services through a platform) (Branchencenter Updates, 2025). Also, instead of making initial public offerings, fintechs preferred mergers or acquisitions by other companies (Branchencenter Updates, 2025).

Building on the discussion of fintech strategies and market adaptation, over 2023 and 2024, fintech funding mainly focused on software-as-a-service ("SaaS"), especially in embedded finance, integration of application programming interfaces ("API") as well as banking-as-a-service ("BaaS") (Finastra, 2023).

Even though there was a recovery at the beginning of 2024 (Germany Trade & Invest, 2025), the economic struggle in the fintech market continued in the second half of 2024, and investments fell to 2018 levels at USD 815 million (KPMG, 2025a, 2025b).

Amid these financial fluctuations, regulatory developments added another layer of change for fintechs. In 2024, important regulations were approved or entered into force that may affect the fintech industry, including EU Instant Payments Regulation ("IPR"); Markets in Crypto-Assets ("MiCA"); and EU AI Act.

Looking ahead, it can be seen that the regulatory focus is increasingly on risk management. For 2025, the German Federal Financial Supervisory Authority ("BaFin") identified six economic risk areas for focus (Risks in BaFin's Focus 2025, 2025). BaFin stressed that the increasing political uncertainty and conflicts also arose in the digital space, with a rising number of cyberattacks (Risks in BaFin's Focus 2025, 2025). It was reported that financial sector companies must be prepared for cyber incidents; otherwise, such incidents may cause a serious threat to financial stability (Risks in BaFin's Focus 2025, 2025). Therefore, the main risks BaFin considers include cyber incidents, outsourcing of IT services, and inadequate measures to prevent money laundering.

Even though the EU economy is considered fragile amid global political conflicts, the Draghi Report from September 9, 2024, and the EU Commission's focus on EU competitiveness may raise support for fintech companies in the EU (Taylor Wessing, 2024). This underscores that, while risks persist, strategic initiatives at the EU level could provide fintechs with new avenues for growth and resilience.

2 Key Fintech Regulations in Germany and the EU

2.1 *Legal Framework in Germany*

Considering that fintech initiatives in Germany are subject to both EU and national legislation, including some discussed in this chapter, numerous laws and regulations which are relevant to them, as explained in the EU legislation section of this chapter.

In this chapter, only recent regulatory developments are discussed, while national legislation is briefly touched on.

Fintech businesses in Germany are regulated based on a number of characteristics, including the type of business they do and what type of contract they facilitate (Krimphove, 2019; Kunz, 2022). Therefore, the entire regulatory framework on licenses and marketing restrictions may be relevant to them (Kunz, 2022).

Primarily, the types of licenses to consider are regulated under the Banking Act (Kreditwesengesetz—KWG); Investment Firm Act (*Wertpapierinstitutsgesetz*—WpIG) that is implementing Directive (EU) 2019/2034 on the prudential supervision of investment firms; the German Electronic Securities Act (*Gesetz zur Einführung elektronischer Wertpapiere*—eWpG); the Payment Services Oversight Act (*Zahlungsdiensteaufsichtsgesetz*—ZAG), that is implementing the Payment Services Directive 2; Capital Investment Act (*Kapitalanlagegesetzbuch*—KAGB) and the Industrial Act (*Gewerbeordnung*—GewO) and the Insurance Supervisory Act (*Versicherungsaufsichtsgesetz*—VAG) (Deutsche Bundesbank, Laws & Regulations, n.d.; Krimphove, 2019; Kunz, 2022).

Regarding the competent authorities to grant licenses, the European Central Bank ("ECB") ranks at the top of the hierarchy being the banking supervisory authority at the European level, granting licenses for banks and systemic investment firms (European Central Bank). At the national level, BaFin covers non-lending financial services, investment, insurance, and payment services (BaFin Website). Following them, the German federal-level authorities would be competent to grant licenses under the *GewO* for certain commercial fintech activities (Kunz, 2022), such as *Bezirksämter* in Berlin and *Industrie- und Handelskammern (IHK)* in Bavaria (*IHK Frankfurt am Main*).

In German law, marketing regulations affecting fintechs are governed by the general principles of fairness, transparency, and non-misleading marketing, as set out in the Act against Unfair Competition (Kunz, 2022; Schroeder, 2008).

For investment services, marketing overlaps with rules under the Securities Trading Act (*Wertpapierhandelsgesetz*—WpHG), the regulation outlining the organisational standards and conduct guidelines for investment services firms (*Verordnung zur Konkretisierung der Verhaltensregeln und Organisationsanforderungen für Wertpapierdienstleistungsunternehmen—WpDVerOV*), as well as the second Markets in Financial Instruments Directive ("MiFID II"), requiring disclosures, i.e., key information sheets (Kunz, 2022). Additional obligations stem from BaFin's Minimum Requirements for the Compliance Function and Additional Requirements governing Rules of Conduct, Organisation and Transparency (*Mindestanforderungen an die*

Compliance-Funktion und weitere Verhaltens-, Organisations- und Transparenzpflichten—MaComp), and the KAGB for marketing of funds (BaFin, n.d.; Kunz, 2022).

For payment services, pre-contractual disclosures are governed by the German Civil Code (*Bürgerliches Gesetzbuch*—BGB) and the Introductory Act to the *BGB*, while marketing securities or financial assets may trigger a requirement for publishing a prospectus under the *WpPG* (Kunz, 2022).

In Germany, payment services providers must deliver detailed pre-contractual information under civil law (Section 312d et ceq. BGB). The BGB and the Introductory Act to the German Civil Code (*Einführungsgesetz zum BGB "EGBGB"*) Article 248 requirements have specific provisions regarding payment service providers (Art. 248 EGBGB; Section 675f et ceq. BGB; Ahmedi, PayTech Law, 2020), while offers of securities to the public typically trigger a formal prospectus obligation under the Securities Prospectus Act (WpPG) within the European capital markets regime (Mock, 2020).

Building on these financial regulations, the draft Financing for the Future Act II (*Zukunftsfinanzierungsgesetz II*—ZuFinG II) removes certain regulatory hurdles under the Investment Act (KAGB) for public investment funds (Cramer et al., 2025). Similarly, the proposed changes to the Investment Act, the tax-exempt maximum for employee share ownership is planned to be increased in Germany to make it a preferred location for start-ups (Die Bundesregierung, 2023).

In a further development concerning digital finance, a noteworthy recent regulatory development in Germany concerning the digitalisation of the financial market was the entry into force, fully as of 30 December 2024, of a framework law, namely the Act on the Digitalisation of the Financial Market (*Gesetz über die Digitalisierung des Finanzmarktes—"FinmadiG"*) (BaFin, n.d.; Federal Gazette, Federal Republic of Germany. 2024; Art 23 of the FinmadiG). FinmadiG enacted the Act on the Supervision of Markets for Crypto-Assets (*Kryptomärkteaufsichtsgesetz—"KMAG"*) as a new standalone act (Art. 1 of the *FinmadiG*). In addition, FinmadiG amended existing legislation such as the Banking Act (*Kreditwesengesetz*), the Securities Trading Act (*Wertpapierhandelsgesetz*) and the Payment Services Supervision Act (*Zahlungsdiensteaufsichtsgesetz*) (BaFin, n.d.; Federal Gazette, Federal Republic of Germany. 2024). As the KMAG is the national implementing legislation of the MiCA Regulation, this step can be considered significant in the German fintech landscape, in particular for crypto service providers.

Complementing these laws, official authorities concerning fintech undertakings in Germany include the Federal Ministry of Finance having departments for fiscal policies, federal budget, tax systems, regulation of financial markets and enhancing digital finance solutions; and the Federal Ministry for Economic Affairs and Climate Action which creates economic policies that promotes sustainable economy and focuses on innovation, digitalisation, competitiveness and sustainability (FinTech Made in Germany, 2024).

Given the vast range of possibilities in which business fintechs operate, many laws may be relevant to fintech initiatives in Germany, in addition to the rules and regulations governing financial services under EU law. Therefore, all legislative frameworks need a detailed analysis. However, since this goes beyond the scope of this report, only a non-exhaustive list of these frameworks, highlighting their most important aspects, is provided.

2.2 *Selected Legal Framework in the European Union*

In the EU legislation, no single legislative framework exists governing all aspects of fintech services. In contrast, fintechs that provide financial services, such as payments and financial advice, are subject to the same laws as other undertakings that provide those services (Stamegna & Karakas, 2019). For instance, payment services are regulated under Directive (EU) 2015/2366 (PSD2) and the proposed Payment Services Regulation (PSR) and PSD3, electronic money under Directive 2009/110/EC, crowdfunding and crypto assets under Directive 2014/65/EU (MiFID II), distance marketing of financial services under Directive 2002/65/EC, and e-commerce under Directive 2000/31/EC (Polasik et al., 2024; Stamegna & Karakas, 2019).

The European Banking Authority identified that non-harmonised laws within the EU can limit fintechs' ability to operate cross-border (European Banking Authority, 2019).

Furthermore, although not directly finance-related, some laws are relevant for fintechs, for instance, the laws on cyber resilience, such as the Digital Operational Resilience Act (Regulation (EU) 2022/2554, "DORA"), which strengthens cybersecurity and IT risk management in financial entities, including crypto-asset providers, by bringing operational obligations such as managing ICT risks and reporting significant incidents (European Securities and Markets Authority, "ESMA" Press Release, 2025). DORA is complemented by the EU Cyber Resilience Act (CRA; Regulation (EU) 2024/2847), which seeks to protect digital products and mitigate cyberattack risks (Parente, 2024; Ruohonen et al., 2025).

European Supervisory Authorities ("ESAs"), which are responsible for the supervision of fintech activities, monitoring any emerging risks for market actors such as users, financial institutions adopt measures and promoting the safety of markets and the consumers in the regulatory field are as follows: the European Banking Authority ("EBA"), the ESMA and the EIOPA (ESMA, 2018).

Some of the laws relevant to fintech undertakings are discussed in this section. In contrast, other legal frameworks, such as DORA, anti-money laundering, and countering the financing of terrorism, are not examined within the scope of this chapter.

For further details on the EU Instant Payments Regulation (EU) 2024/886 (adopted on March 13, 2024), Chapter 5.

2.2.1 Payment Services Directive 2 (Directive (EU) 2015/2366—PSD2) and the Forthcoming Payment Services Directive 3 ("PSD3") & Payment Services Regulation ("PSR")

PSD2 and PSD3 are highly relevant for fintech initiatives in Germany, as payment services were among the first sectors in which German fintech companies began operating (Noerr, 2022).

The PSD2 was implemented into German law in January 2018 (Deutsche Bundesbank). In 2020, the Commission on a Retail Payments Strategy for the EU published that the PSD2 would be reviewed to determine whether it continues to be relevant to its purpose, taking into account the latest market developments (European Commission, 2023).

One of the most noteworthy effects of PSD2 was that banks were mandated to provide access to their data and applications via APIs, accelerating the adoption of the open banking concept (Şahin & Cantürk, 2020). In this context, open banking initiatives expanded, driven by fintech applications that transformed the banking sector into an interconnected financial ecosystem (Şahin & Cantürk, 2020).

While PSD2 was a significant step towards the development of the open banking sector, Germany had already been using APIs in financial applications, such as credit notes, even before PSD2, and therefore successfully adapted to open banking applications (Şahin & Cantürk, 2020).

In 2023, under the EU Commission's Payments Package, it was proposed to amend PSD2 and PSD3 and to draft the first Payment Services Regulation (PSR). The EU Commission announced measures on six key points to modernise PSD2 and aiming to: (1) Combat and mitigate payment fraud such as extending refund rights of victims, (2) improve consumer rights by providing more transparent information, (3) enhance the functioning of open banking, (4) increase the availability of cash in stores and through ATMs (5) bolster harmonisation and enforcement and (6) enhancing the competitive position of non-bank payment service providers compared to banks by granting them access to all EU payment systems (European Commissio, 2023).

On June 13, 2025, the Council of the EU released compromise texts for the PSD3 and the PSR and tasked the Presidency with negotiating an agreement with the European Parliament (Council of the European Union, 2023).

The PSD3 is expected to address challenges, such as the increased need for security and efficiency in digital payments, given changes in payment practices, especially after the COVID-19 pandemic (Recital 1, Ordinary Legislative Procedure (COD) 2023/0209). PSD3 and the PSR framework are expected to benefit both consumers and fintechs, as the rules will be more straightforward for both parties, consumer rights will be enhanced, and market entry barriers to payment services will be removed (Škrabka, 2024).

PSD2 applies to payment service providers, which are classified into three primary categories: credit institutions, payment institutions ("PIs"), and electronic money institutions ("EMIs") (Herrera, 2025). Under the proposed PSD3, the distinction between PIs and EMIs will disappear, and a single PIs category will be created, with EMIs classified as PIs offering electronic money services (Herrera, 2025).

The draft PSR aims to prevent forum shopping by providing a harmonised application within the EU, as it will be directly applicable in their legislative systems. At the same time, PSD3 broadens consumer protection and fair competition between banks and non-banks (Škrabka, 2024).

To address unfair competition and respond to complaints from market actors, the right for payment service providers ("PSPs") to have direct access to payment systems rather than indirect access is to be introduced under PSD3 (Škrabka, 2024).

The draft PSR includes detailed provisions to combat fraud and protect customers, enhancing users' rights. Chapter 7 of the PSR aims to achieve the goals of combating fraud and protecting customer rights by regulating issues such as fraud reporting, transaction monitoring, payment service providers' data processing, strong customer authentication ("SCA"), and security measures to protect users (PSR Proposal).

It can be expressed that the improvements to be brought to SCA under the PSR proposal contribute to sustainable development goals in the EU Member States since PSPs are mandated to bring effective methods of enabling SCAs for users with disabilities, older people, or other vulnerable users who lack access to payment methods or digital platforms (Art. 88 PSR Proposal).

Also, in Chapter 4 of the PSR, provisions are being introduced regarding the responsibilities of payment service providers, such as providing a free-of-charge verification service that verifies and communicates whether a match between the unique identifier and the payee's name eventuates. The PSPs are held liable for financial losses arising from their failure to notify and from discrepancies between the IBAN and the payee's name. The PSPs also have an obligation to refund the full amount to users who are victims of fraud if the users are unlawfully manipulated by fraudsters using the PSPs' contact information.

The German Banking Industry Committee published its comments on the upcoming Payment Services Regulation regarding its regulatory fields, such as fraud prevention, third-party providers, strong customer authentication, and its relation to MiCA (The German Banking Industry Committee, 2025). They noted that while protecting consumers is crucial, the liability for impersonation fraud imposed on PSRs is too broad, potentially harming PSPs even when the fraud occurs outside their influence and detectability (The German Banking Industry Committee, 2025). It can be argued that, since fintechs are more vulnerable than traditional banks, the new liability may pose a greater challenge for them.

Regarding open banking, an obligation is planned to be introduced for account servicing payment service providers ("ASPSPs"), requiring them to supply users permission dashboards to behold and control data access to open banking services (Recital 65 of the PSR Proposal; Art. 43 of the PSR Proposal). This can be interpreted as PSR aims to address data protection challenges and non-compliance with the GDPR that were present in PSD2 (Škrabka, 2024).

2.2.2 The Financial Data Access Regulation ("FiDA")

The EU Commission also proposed a legislative framework to clearly define rights and responsibilities for controlling the financial sector's sharing of client data (European Commission, 2023). The data in question is not limited to payment accounts. FiDA complements the proposed PSR by enabling access to financial data beyond payment accounts, to encourage data-driven innovation in financial services (Council of the EU, Press release, December 4, 2024).

The EU Commission set out six key points under the proposal: (i) customers have the freedom to choose whether to share their data with users in a machine-readable format to receive better services; (ii) data holders are responsible for enabling data sharing with data users while considering the necessary technical measures and consumer permissions; (iii) customers get increased control over who accesses their data and for what purposes, in line with the GDPR; (iv) the proposal supports the standardisation of certain aspects, such as the required technical interfaces; (v) it introduces clear liability regimes for data incidents and dispute resolution mechanisms; (vi) it offers incentives for data holders to create high-quality interfaces for data users (European Commission, 2023).

FiDA is an EU initiative that aims to empower consumers by giving them increased control over their financial data. It facilitates secure financial information access, thereby enabling consumers to share their data with authorised third parties to improve financial services, such as personalised advice and product recommendations (European Commission, 2023). FiDA improves transparency and innovation while ensuring high data protection standards (European Commission, 2023). The FiDA provisions concerning financial data access control panels are expected to enhance customer confidence by strengthening users' control over how their data is shared (Sempere Pastor, 2025).

2.2.3 The Regulation on Harmonised Rules on Fair Access to and Use of Data (Regulation (EU) 2023/285—"Data Act")

Became applicable on September 12, 2025, the Data Act underlines the importance of the interoperability of the data for achieving economic benefits such as sustainable growth and competitiveness, and highlights the necessity of a harmonised framework within the Union to enable an internal data market which addresses the needs of data-driven digital economy (Recitals 1-5 and Art. 50 of the Data Act; The European Commission, 2025c).

Building on this objective, the Data Act concerns data flows between businesses to other actors, including users of the services, data subjects, other businesses, public sector bodies and other bodies. By simplifying the data flows, it aims for connected devices to share data between them, give users the availability of wider options of services, and the possibility to obtain information regarding the data of users which is generated in connection with the usage of a product (Recital 5 of the Data Act; The European Commission, 2025c).

Moreover, the Data Act makes it easier to use parallel use of providers and services, which requires ongoing transfer of data (Recital 99 of the Data Act). The Data Act allows services to charge for the ongoing transfers, provided that they are not more than actual and objective costs, meaning that they cannot profit over such transfer charges (Art. 34 & Recital 99 of the Data Act). Parallel use could be important for services since different providers' services may be complementing each other or provide a stronger IT security in cases where one of the providers is unavailable. The parallel use may be particularly relevant for fintechs that adopt multi-cloud strategies to increase resilience and ensure compliance with regulations such as DORA, by avoiding reliance on a single provider (Recital 99 of the Data Act).

Against this background, existing industry research confirms that cloud data egress costs can create significant cost barriers to multi-cloud use (Krishnakumar, 2025; Westfall & Dickens, 2025). Accordingly, the implementation of the Data Act, by limiting such charges to cost recovery, may provide meaningful relief for digital businesses, including fintechs.

In a similar vein, the Data Governance Act (Regulation (EU) 2022/868) (entered into force in June 2022 and applicable since September 2023) may be considered as relevant for fintechs, since it strives for data availability and aims to enhance Common European Data Spaces in fields including finance (The European Commission, 2024). While distinct from the Data Act, it complements the broader EU data strategy by promoting trusted mechanisms for data sharing and reuse, and consequently, supporting the data-driven environment in which fintechs operate.

2.2.4 Markets in Crypto-Assets Regulation (Regulation (EU) 2023/1114—"MiCA", Entered into Force on June 29, 2023, Fully Applicable as of December 30, 2024)

MiCA establishes licensing for crypto asset service providers (CASPs), issuers of asset-referenced tokens (ARTs), and issuers of electronic money tokens (EMTs), regulates the issuance and admission to trading of cryptoassets, clarifies the obligations applying to them, and strengthens consumer protection. At the same time, certain activities remain outside its scope: non-fungible tokens (NFTs), fully decentralised services, and certain assets covered by other financial law are exempted from the scope of MiCA (Zetzsche & Woxholth, 2025).

This regulation emerged as a response to previously fragmented legislation within the EU regarding crypto-assets and aims to ensure uniform conditions for firms in the crypto-asset market (MiCA Explanatory Memorandum, 2).

More specifically, MiCA seeks to enable CASPs to leverage their businesses on international level and provide access to banking services for them to run their businesses without obstacles, simplifying and harmonising the making market entry for the issuers of crypto assets and CASPs, to support financial stability prevent risks from stablecoins, while avoiding unnecessary regulation of the underlying technology to keep the EU globally competitive (Recital 6, MiCA). In line with this objective, Recital 8 of the MiCA suggests that the EU should aim for a convergent approach regarding crypto assets and CASPs.

Within this harmonised framework, the passporting concept introduced by MiCA is expected to benefit fintechs, as companies with an EU license can operate across all Member States without having to comply with different national legislations (Mkrtchyan & Treiblmaier, 2025). MiCA is important for fintechs because it opens the door for them to develop regulated stablecoins, tokenised assets, and DeFi platforms that are safely integrated into the financial system.

Nevertheless, some challenges arising from the application of MiCA could include its overlay with other legal frameworks, i.e., MiFID II, in correctly classifying which regulation a crypto-asset falls under (Martínez Nadal et al., 2025; Mkrtchyan & Treiblmaier, 2025). It can be particularly challenging for companies providing both traditional financial services and crypto-asset services to navigate the two legal frameworks (Mkrtchyan & Treiblmaier, 2025).

In this context, the relationship between MiCA and MiFID II is further nuanced by the exemption granted to institutions already authorised under MiFID II; such institutions are not obligated to obtain a separate MiCAR licence to provide crypto-asset services and can continue offering such services under their existing authorisation (Hanten et al., 2022). This exemption may encourage new market participants to obtain MiFID II authorisation initially to avoid licensing obligations (Hanten et al., 2022).

2.2.5 MiFID II (Directive 2014/65/EU on Markets in Financial Instruments)

Entered into force in January 2018, MiFID II was transposed in a fragmented manner across the EU Member States, as it does not provide a single definition covering all financial instrument types (Echebarría Sáenz, 2025; Tomasz, 2022).

A list of financial instruments was given in Annex I, Section C, of MiFID II, including transferable securities, money-market instruments, units in collective investment undertakings, various derivative contracts (e.g., options, futures, swaps), and emission allowances, rather than outlining general conditions or criteria (Echebarría Sáenz, 2025).

Therefore, one Member State may classify an asset as a financial instrument, thereby excluding it from the scope of MiCA. At the same time, the other Member States take a different approach, which may result in crypto asset issuers seeking EU jurisdictions where they are not subject to MiFID II, potentially more cumbersome (Tomasz, 2022).

Despite this, MiFID II has a significant impact on bank servicing investors, as it aims to strengthen investor protection (Loonen & Janssen, 2023).

2.3 *Other Relevant Legislative Frameworks*

2.3.1 GDPR (Regulation (EU) 2016/679—General Data Protection Regulation)

GDPR applies to all data controllers or processors whose processing activities affect individuals, irrespective of their location, either by offering goods or services to them or by monitoring their behaviour in the EU (Art. 3 GDPR). Given the territorial scope of the GDPR and the broad definition of personal data, fintechs within the EU, or those activities relating to data subjects in the EU, are subject to this regulation (Art. 3 & 4 (1) GDPR). Serious fines of up to 4% of an organization's global yearly sales or 20 million euros may be imposed for noncompliance with the GDPR (Art. 83 GDPR). Therefore, fintech initiatives must put processes in place to comply with the GDPR and continuously monitor their compliance.

In addition, the GDPR is relevant to fintechs considering its relation with PSD2. It should be noted that Article 94 of the PSD2 states that data processing for PSD2 has to be conducted in compliance with the GDPR and the data protection rules implementing Regulation (EU) 2018/1725. PSD2 introduced more specific requirements regarding data protection for payment service providers, including the prohibition on sharing users' banking data with third parties and the requirement to rely on explicit consent (Art. 94(2) PSD2).

It is noteworthy that Art. 94(2) of PSD2 mandates that PSPs obtain "explicit consent". Under the GDPR, explicit consent must be given by an express statement for specific processing purposes, usually in writing, and is one of the exceptions to the prohibition on processing sensitive data under Article 9 GDPR. To process sensitive data, there must be a legal basis under Art. 6 GDPR in addition to explicit consent. However, the EDPB stated that the concepts of explicit consent under PSD2 and GDPR differ (European Data Protection Board, 2020). In principle, the PSPs need to process data to perform a contract under Art 6(1)(b) GDPR, and Art 94(2) PSD2 cannot be evaluated as an additional legal basis for the processing (European Data Protection Board, 2020). Instead, explicit consent should be regarded as an additional contractual requirement for access to, subsequent processing, and storage of personal data (European Data Protection Board, 2020).

Particular academics criticised the differences in the meaning of explicit consent and the misalignments between the two legal texts (Ferretti, 2022; Gounari et al., 2024).

As mentioned above, PSR is expected to address specific data protection challenges and non-compliance with the GDPR (Škrabka, 2024). However, there are points to be clarified in the proposals of the PSD3 and PSR according to the opinion of the European Data Protection Supervisor (EDPS, 2023).

In the EDPS opinion on the PSD3 and PSR proposals, the wording of Recital 69 of the PSR proposal was discussed. Recital 69 of the proposal of the PSR reads "(..) permission should not be construed exclusively as 'consent' or 'explicit consent' as defined in Regulation (EU) 2016/679". The EDPS criticised the use of "exclusively" and stated that it creates uncertainty. It does not serve to differentiate between the "permission" which refers to the acceptance of the commercial service by the user and the consent referred to in Art 6(1)(a) and Art 9(2)(a) GDPR (EDPS, 2023). Therefore, it is recommended that Recital 69 be amended as "permission should not be construed as 'consent' or 'explicit consent' or 'necessity for the performance of a contract' as defined in Regulation (EU) 2016/679" (EDPS, 2023).

It can be seen that EDPS aims to clarify the legal basis for data processing for PSR and to avoid the misunderstanding that the processing is based on consent, as discussed in the EDPB Guideline.

The EDPS opinion noted that financial data can reveal sensitive information about a data subject, and it is welcome that the proposals require ASPSPs to provide users with a dashboard.

To reduce the risk of personal data being shared with unauthorised parties by ASPSPs, the EDPS recommended that dashboards should refer to the specific payment services for which permission has been granted, that access by third-party services should be limited in line with data minimisation principles, and that the legal basis for each access request should be clearly defined. Furthermore, the EDPS recommended allowing ASPSPs to verify the permissions granted by users or to implement sufficient alternative safeguards.

Lastly, while aligning with the GDPR is on the table in relation to financial applications, it is worth noting that the EU Commission proposed to extend certain rules in the GDPR, such as allowing AI systems to process sensitive personal data to prevent bias, which may find application as part of fintech services (The Digital Omnibus package, 2025).

2.3.2 The EU AI Act (Regulation (EU) 2024/1689)

Applicable as of August 2, 2026, the EU AI Act regulates AI systems based on risk levels, introducing restrictions on high-risk systems (e.g., transparency, accountability, human oversight). However, threre is a phase by phase entry into force depending on the Chapeters of the AI Act: The enforcement rules of the AI Act (Chapter 9) will apply as of August 2, 2026, while General Provisions (i.e., AI literacy) and Prohibited AI Practices apply from 2 February 2025, certain provisions such as the ones on High-Risk AI Systems apply from 2 August 2025, and specific obligations such as Article 6(1) related to high-risk AI categories will apply from 2 August 2027 (Regulation (EU) 2024/1689, Art. 113). Despite the rules on the entry into force in the AI Act, the recent Digital Omnibus package proposes to extend the deadlines, considering the feedback from the industry and to ease the implementation challenges (AI Act Service Desk FAQ, n.d.; European Commission, 2025a). For instance, the implementation of the rules regarding high-risk AI systems is proposed to be extended up to 16 months(AI Act Service Desk FAQ, n.d.; European Commission, 2025b).

Given the rising trends of AI use in their operations (KPMG, 2025a, 2025b), the EU AI Act is highly relevant to fintech initiatives in Europe. AI finds applications in various fields of finance, including financial risk management and advisory services, lending, trading, and fraud detection (Ozili, 2025). Under the AI Act, these applications must be assessed in light of their potential impact on individuals' fundamental rights in order to determine their risk classification. For instance, since credit scoring decisions may affect the use of essential private or public services or benefits, such as access to loans or electricity, this is explicitly mentioned in the AI Act as a High-Risk AI category.

Under Recital 58 of the AI Act, AI systems used for credit scoring may lead to or create new forms of discrimination between persons (Recital 58 AI Act). Similarly, the Recital 58 also notes that AI systems used to assess risk and set pricing for health and life insurance are classified as high-risk since they can significantly affect individuals' lives. The AI Act highlights that, unless carefully designed, developed, and implemented, they may result in serious consequences, such as financial exclusion (Recital 58 and Annex III: High-Risk AI Systems Referred to in Article 6 (2) point (5) (c) AI Act).

By contrast, certain AI applications commonly used by fintechs, such as financial fraud detection is exempted from this classification (Annex III: High-Risk AI Systems Referred to in Article 6 (2) point (5) (b) AI Act).

It seems that the AI Act acknowledges the interplay between AI governance and Union financial services law. For instance, the AI Act specifies that compliance is primarily achieved through existing EU financial services laws regarding quality management systems (Art. 17, para 4), technical documentation (Art. 17, para 3), and documentation retention (Art. 18 (3), automatically generated logs (Art. 19, paras 1–2), monitoring and log-keeping obligations for deployers of high-risk AI systems (Art. 26, para 5).

This integration is particularly evident in relation to the post-market monitoring obligation for high-risk AI systems. Under Article 72 of the AI Act, providers of high-risk AI systems are required to establish post-market monitoring systems. However, where such providers are financial institutions, the AI Act allows the necessary obligations to be incorporated into their systems and plans under already existing Union financial services law, so that they can avoid duplications and burdens, with the condition of providing an equivalent level of protection (Art. 72 (4) AI Act). Therefore, it can be concluded that fintechs offering services such as credit scoring and pricing must pay attention to the obligations of deployers and providers of high-risk AI systems and ensure that their existing operations achieve the level of protection under the AI Act.

While the AI Act does not expressly specify what is included in the Union financial services law, it can be interpreted to encompass all laws relating to financial services. In fact, in an article published by the EIOPA, it was stated that AI applications used by financial institutions, including outsourced tools and services, can be effectively managed under existing EU legislation, particularly the Digital Operational Resilience Act (DORA), which covers "critical third-party service providers" (Parente, 2024). Beyond these critical services, most AI use cases in areas such as claims management, anti-money laundering, or fraud detection are already regulated under current financial laws, without the need to add additional obligations

(Parente, 2024). It is, however, important to assess whether existing rules sufficiently address AI-specific risks or whether additional guidance is needed on issues such as proportionality, fairness, explainability, and accountability (Parente, 2024).

2.3.3 EU Digital Identity Wallet (Regulation (EU) 2024/1183—eIDAS2)

eIDAS2 builds on eIDAS by mandating that EU Member States provide interoperable digital ID wallets by November 2026 for secure sharing of identity credentials. It enables streamlined KYC onboarding for financial institutions via trusted credential exchange in a standardised, secure infrastructure. It may be vital for embedded finance, lending, payments, and insurtech services.

3 Overview of Fintech Regulations in Türkiye

3.1 Fintech Ecosystem in Türkiye

Various sources cite the COVID-19 pandemic and Türkiye's young population as reasons for the swift growth of fintech applications in the country (Ovenc & Nabiyev, 2025; Onay et al., 2024).

Indeed, according to recent statistics, Türkiye has the youngest median age, at 34 (TUIK, 2025).

According to a report prepared by the Presidency of the Republic of Türkiye Finance Office, the number of Licensed E-Money and Payment Institutions in Türkiye has kept growing since 2018 (The State of Fintech Ecosystem in Türkiye, 2024). As of October 2025, there are 63 e-money institutions and 27 payment institutions (TCMB).

While the number of fintech institutions in Türkiye reached 916 in 2025, the most prominent fintech verticals are respectively payment technologies, banking technologies, and blockchain/crypto technologies (FinTech İstanbul, 2025; T.C. Cumhurbaşkanlığı Finans Ofisi, 2025).

The Banks Association of Türkiye published that as of December 2024, the number of active retail digital banking customers reached 114.4 million (Banks Association of Türkiye, 2025).

Considering the fintech ecosystem and regulations in Türkiye, this sub-chapter focuses on regulations governing Payment Services, EMIs, digital banking, and BaaS.

3.2 *Selected Regulations in Türkiye*

3.2.1 The Law No. 6493 on Payment and Securities Settlement Systems, Payment Services, and Electronic Money Institutions

In Türkiye, Electronic Money Institutions and Payment Institutions are regulated under Law No. 6493 and, as of January 1, 2020, are supervised and audited by the TCMB (The Central Bank of the Republic of Türkiye). The reference regulations for Law No. 6493 have been the Payment Services Directives 1 and 2 (Şahin & Cantürk, 2020).

The secondary legislation derived from Law No. 6493, which relates to E-money and payment institutions, is as follows (TCMB): (i) Regulation on Payment Services and Electronic Money Issuance, Payment Service Providers, (ii) Regulation on the Generation and Use of TR QR Code in Payment Services, (iii) Regulation on the Disuse of Crypto Assets in Payments and (iv) Communique on the Management and Supervision of the IT Systems of Payment and Electronic Money Institutions and the Data Sharing Services of Payment Service Providers in the Payment Services Area (TCMB).

3.2.2 Regulation on Payment Services (December 1, 2021, OG 31676)

It governs the authorisation and operation of PIs and EMIs, including the provision of payment services to other providers and the issuance of electronic money (CBRT, 2021; Hanten et al., 2022 Annual Report).

3.2.3 QR Code Regulation (August 21, 2020, OG 31220)

It establishes rules for payment transactions executed via QR codes under the Law on Payment Systems, specifying operational procedures and technical requirements (Hanten et al., 2022).

3.2.4 Non-Use of Crypto Asset Regulation (April 30, 2021)

It prohibits the use of crypto-assets for payments (Hanten et al., 2022; Sönmez et al., 2024).

3.2.5 Amendments to the Capital Markets Law No. 6362 Regarding Crypto Assets (Law No. 7518) (July 2, 2024, OG 32590)

Law No. 7518 introduces the first comprehensive statutory framework for crypto assets in Turkish financial law (Cindemir, 2025). With this amendment, CASPs, including entities engaged in the regular commercial activities of crypto asset trading, exchange, transfer, custody, and management of private keys, are brought within the scope of the Turkish Capital Markets Board ("SPK") regulatory and supervisory authority (Law No. 7518, Art. 2 et seq.). On **March 13, 2025**, the secondary legislation was published in the *Official Gazette* (No. 32840), relying on Articles 35/B and 35/C of Law No.7518. These are (i) Communiqué III-35/B.1, which regulates the establishment and operations of CASPs, including the procedures and principles related to their information systems and technological infrastructure (Communiqué III-35/B.1, Art 1 & 2); and (ii) Communiqué III-35/B.2 which sets out rules regarding the services and activities that CASPs may offer, particularly with respect to crypto asset trading platforms, the transfer of crypto assets, and the listing of such assets (Communiqué III-35/B.2, Art. 5 et seq.).

3.2.6 The Amendment to the Anti-Money Laundering Regulation (May 1, 2021)

It extends AML/CFT obligations to crypto-asset service providers (Hanten et al., 2022).

In addition to the TCMB, the key institutions which audit e-money and payment service providers are as follows: (i) The Financial Crimes Investigation Board (MASAK), which is a main service unit of the Ministry of Finance and is directly attached to the Ministry of Finance, in accordance with the MASAK legislation, (ii) The Ministry of Trade, under Law No. 6563 on the Regulation of Electronic Commerce, the Consumer Protection Law No. 6502, and the Turkish Commercial Code No. 6102, especially the provisions regarding joint stock companies, (iii) The Banking Regulation and Supervision Authority (BDDK) (*eng.* BRSA), in terms of POS usage, and (iv) The Ministry of Treasury and Finance and the Revenue Administration, in terms of taxation (Perçin, 2023, Sorularla Fintek).

3.2.7 Legislation Related to Digital Banking & Banking as a Service

The supervisory authority responsible for digital banking is the Banking Regulation and Supervision Authority (BRSA).

According to the BRSA, there are five digital banks, of which two of are participation banks (A type of interest-free bank operating under Islamic finance rules [Arzova & Şahin, 2025]) and three are deposit banks as of October 2025 (BRSA Website).

The main legislation related to the Digital Banking & BaaS is as follows (Perçin, 2023, Sorularla Fintek); (i) The Banking Law No. 5411), (ii) Regulation on the Operating Principles of Digital Banks and Banking as a Service Model (Digital Banking Regulation), (iii) Regulation on Remote Identification Methods to be Used by Banks and the Establishment of Contract Relationship in the Electronic Environment, (iv) The Regulation on Indirect Shareholding and Transactions Subject to Permission of Banks, (v) Regulation on Information Systems and Electronic Banking Services of Banks, (vi) Regulation on Measurement and Assessment of Capital Adequacy of Banks, (vii) Regulation on The Disclosure of Confidential Information, (viii) Additional Explanations on The Implementation of The Regulation on The Disclosure of Confidential Information, (ix) The Regulation on the Procurement of Support Services by Banks, and (x) The Communiqué on the Procedures and Principles Regarding the Fees to be Received from Financial Consumers (Bankacılık Düzenleme ve Denetleme Kurumu).

3.2.8 Regulation on the Operating Principles of Digital Banks and Banking as a Service Model (Digital Banking Regulation)

In Turkish legislation, digital banking is regulated by the regulation on the Operating Principles of Digital Banks and BaaS Model (Digital Banking Regulation), published in the Official Gazette numbered 31704 and dated December 29, 2021, drafted by BRSA and entered into force on January 1, 2022.

However, the digital banks are also subject to all regulations that apply to credit institutions (Article 4 (1) of the Digital Banking Regulation), particularly the Banking Law No. 5411, which was the basis for the Digital Banking Regulation (Hanten et al., 2022). Therefore, the legislation relates to digital banking, and traditional banks are similar (Perçin, 2023, Sorularla Fintek).

It is noteworthy that there is an activity restriction for digital banks under Article 5 (1) of the Digital Banking Regulation, which is that credit customers of digital banks can only consist of financial consumers and SMEs.

It can be suggested that this limitation contributes to the financial inclusion of smaller players in the market. Moreover, Article 11 of the abovementioned Regulation promotes financial inclusion by requiring applicants seeking to establish a digital bank to include, within their business plans, an analysis of the needs of their target audience, such as students, housewives, individuals under eighteen, and SMEs, and to specify the products and services designed to address these needs and enhance financial inclusion.

As exceptions to the limitation of the customer portfolio, digital banks are permitted to engage in interbank and capital market transactions, extend loans to other banks, and provide foreign currency loans to enterprises exceeding the SME threshold. Furthermore, digital banks may accept deposits from other banks, offer protection account services for payment and e-money institutions licensed under Law No. 6493, and provide BaaS to interface developers that are larger than SMEs (Article 5 (2) of the Digital Banking Regulation).

3.2.9 Regulation on the Information Systems (OG 31069, 15 March 2020; Effective 20 June 2020)

Issued by the BRSA, this regulation governs banks' information systems management, cybersecurity controls, and data localisation, ensuring that digital banking infrastructures and FinTech integrations operate securely within Türkiye (Hanten et al., 2022; Yıldırım, 2024).

3.2.10 Regulation on Remote Identification (OG 31441, 1 April 2021)

It establishes procedures for remote customer identification, enabling banks and fintechs to onboard clients digitally and form contractual relationships electronically, thereby supporting fully online financial services (Hanten et al., 2022; Yıldırım, 2024).

3.2.11 Regulation on Indirect Shareholding and Transactions Subject to Permission of Banks (OG 26333, 1 November 2006)

It is referenced for licensing and ownership conditions applicable to digital banks, ensuring regulatory consistency as technology-based firms enter the sector (Hanten et al., 2022).

3.2.12 Regulation on the Disclosure of Confidential Information (OG 31501, 4 June 2021)

It aligns banking secrecy rules with data protection law, setting strict confidentiality standards for banks and third-party interface providers handling customer data within FinTech partnerships (Gürkaynak et al., 2021; Hanten et al., 2022).

3.2.13 Regulation on the Procurement of Support Services by Banks (OG 28106, 5 November 2011)

It defines the conditions under which banks may outsource support services, clarifying the regulatory treatment of "interface providers" and "service banks" in Banking-as-a-Service (BaaS) models (Hanten et al., 2022).

4 Conclusion

The lack of legal clarity poses a challenge in the age of rapid technological innovation, where laws often struggle to keep pace with emerging applications, including fintech activities such as crypto-asset issuance, digital lending, and AI-based financial services (Arner et al., 2017). This fragmentation can create compliance burdens and uncertainty for fintechs and SMEs, limiting innovation and access to finance.

Nonetheless, recent EU initiatives indicate growing awareness of these challenges and a deliberate move toward regulatory harmonisation. The MiCA introduces a uniform framework for crypto-asset issuance and service provision, enhancing legal certainty and consumer protection across Member States. Similarly, the AI Act allows financial institutions to rely on existing governance, documentation, and monitoring obligations under Union financial services law to satisfy certain AI Act requirements (Articles 17–19), reducing regulatory overlap and compliance duplication.

Similarly, to enhance consistency with the GDPR, the proposed PSR and PSD3 include provisions that aim to harmonise data protection and financial rules by resolving discrepancies and providing clearer distinctions for terms that are used differently across the frameworks.

Furthermore, the FiDA aims to deepen open banking by strengthening consumer rights, standardising data-sharing obligations, and broadening access to financial data beyond payments.

Also, at the national level, initiatives like the Financing for the Future Act II aim to improve access to capital and simplify regulatory hurdles for fintech start-ups in Germany.

In Türkiye, Article 5(1) of the Digital Banking Regulation limits digital banks' credit customers to financial consumers and SMEs, while Article 11 requires consideration of the needs of underrepresented groups. These measures foster access to finance, supporting the SDGs by promoting financial inclusion and reducing inequalities.

It is possible to observe the influence of recent criticisms regarding Europe's overly complex regulatory framework, which has been viewed as a hurdle to innovation (Draghi Report, 2024). In particular, several of the latest reforms appear to reflect suggestions made in the Draghi Report, calling for a more streamlined and innovation-friendly approach across the EU (Draghi Report, 2024). The proposal of the Digital Omnibus package can also be seen as a reflection of the efforts to simplify the rules for sector players.

On the other hand, legal frameworks such as CSRD will bring additional obligations for fintech initiatives in the EU. The regulatory frameworks in the EU remain complex, and they require significant time, resources, and labour for fintech undertakings to comply with. As a recommendation to ease fintechs' burden, fintech hubs supported by the government, such as BaFin's Innovation Hub, may give guidance on broader regulatory challenges in addition to financial regulations and licensing requirements.

While it is clear that a wide range of new regulations is shaping the future of finance in Germany, Türkiye, and the EU, a conscious effort to ease compliance burdens for tech-driven companies may be seen. Whether this shift will empower fintechs or lead to further regulatory fragmentation remains to be seen.

References

Arner, D. W., Barberis, J., & Buckley, R. P. (2017). FinTech, RegTech, and the reconceptualization of financial regulation. *Northwestern Journal of International Law & Business, 37*(3), 371–414. https://scholarlycommons.law.northwestern.edu/njilb/vol37/iss3/2

Arzova, S. B., & Şahin, B. Ş. (2025). Performance comparison of participation banks and deposit banks: The impact of high deposit rates on profitability, liquidity and non-performing loans. *Marmara Üniversitesi Öneri Dergisi, 20*(64), 335–350. https://doi.org/10.14783/maruoneri.1660777

BaFin. FinTech Innovation Hub. https://www.bafin.de/EN/Aufsicht/FinTech/fintech_node_en.html

BaFin. (2025). *Risks in BaFin's Focus 2025*. https://www.bafin.de/DE/Aufsicht/Fokusrisiken/Fokusrisiken_2025/Druckansicht/Fokusrisiken_2025_druck_node.html

CBRT. *The official website of the CBRT* (The Central Bank of the Republic of Türkiye). https://www.tcmb.gov.tr/

CBRT. (2021). *Annual Report*. https://www3.tcmb.gov.tr/yillikrapor/2021/en/m-2-5.html

Cindemir, G. (2025). Avrupa Birliği ve Türkiye'de Kripto Varlık Hukuku: MiCA ile 7518 Sayılı Kanun'un Tanımlamalar Açısından Karşılaştırılması. *Abant Sosyal Bilimler Dergisi, 25*(2), 1144–1154. https://doi.org/10.11616/asbi.1658422

Council of the European Union. (2023). Proposal for a directive of the European Parliament and of the Council on payment services and electronic money services in the internal market (Procedure 2023/0209 (COD)) Document ST 10176 2025 INIT]. https://data.consilium.europa.eu/doc/document/ST-10176-2025-INIT/en/pdf

Cramer, J., Gaffron, F. G., Heuking Kühn Lüer Wojtek. (2025). *Future Financing Act II—Promotion of investments in renewable energies*. https://www.heuking.de/en/news-events/newsletter-articles/detail/zukunftsfinanzierungsgesetzes-ii-foerderung-von-investitionen-in-erneuerbare-energien.html

Deutsche Bundesbank. PSD2: Payment services. https://www.bundesbank.de/en/tasks/payment-systems/psd2/psd2-775954

Die Bundesregierung. (2023). *Key impetus for new growth*. https://www.bundesregierung.de/breg-de/service/archiv-bundesregierung/financing-for-the-future-act-2247606

Directive 2014/65/EU of the European Parliament and of the Council. (2014). On markets in financial instruments (MiFID II) and amending Directives 2002/92/EC and 2011/61/EU (recast). *Official Journal of the European Union, L 173*, 349–496.

Directive (EU) 2015/2366 of the European Parliament and of the Council. (2015). On payment services in the internal market (PSD2). *Official Journal of the European Union, L 337*, 35–127.

Draghi, M. (2024). *The future of European competitiveness—A competitiveness strategy for Europe*. European Commission. https://commission.europa.eu/topics/eu-competitiveness/draghi-report_en

Echebarría Sáenz, M. (2025). Regulating market abuse in crypto-assets. Ca*pital Mar-kets Law Journal, 17(3)*, 365–385.

EDPB—European Data Protection Board. (2020). *Guidelines on PSD2 after public consultation.* https://www.edpb.europa.eu/sites/default/files/files/file1/edpb_guidelines_202006_psd2_afterpublicconsultation_en.pdf

European Banking Authority. (2019, July 19). Report on the regulatory perimeter, regulatory status and authorisation approaches in relation to FinTech activities. https://www.eba.europa.eu/publications-and-media/press-releases/eba-publishes-report-regulatory-perimeter-regulatory-status-and-authorisation-approaches-relation

European Data Protection Supervisor. (2023). Opinion 39/2023 on the Proposal for a Regulation on payment services in the internal market and the Proposal for a Directive on payment services and electronic money services in the Internal Market. *European Data Protection Supervisor*. https://www.edps.europa.eu/data-protection/our-work/publications/opinions/2023-08-22-edps-opinion-392023-regulation-payment-services-internal-market-and-directive-payment-services-and-electronic-money-services-internal-market

European Commission. (2023). *Modernising payment services and opening financial services data: New opportunities for consumers and businesses.* https://ec.europa.eu/commission/presscorner/detail/en/ip_23_3543

European Commission. (2024). *European Data Governance Act.* https://digital-strategy.ec.europa.eu/en/policies/data-governance-act

European Commission. (2025a). *AI Act Service Desk—Frequently asked questions: Timing and application of the AI Act.* https://ai-act-service-desk.ec.europa.eu/en/faq

European Commission. (2025b). *Proposal for a Digital Omnibus Regulation: Simplifying compliance under EU digital legislation.* https://digital-strategy.ec.europa.eu/en/library/digital-omnibus-regulation-proposal

European Commission. (2025c). *Data Act.* https://digital-strategy.ec.europa.eu/en/policies/data-act.

European Parliamentary Research Service (EPRS). Stamegna, C., & Karakas, C. (2019, February). EPRS briefing.

European Securities and Markets Authority (ESMA). (2018, March). *Joint Committee final report on big data* (Report No. JC/2018/04). https://www.esma.europa.eu/sites/default/files/library/jc-2018-04_joint_committee_final_report_on_big_data.pdf

European Securities and Markets Authority (ESMA). *DORA oversight.* https://www.esma.europa.eu/dora-oversight

European Union. (2024). Regulation (EU) 2024/1183. Official Journal of the European Union, L 1183.

EUR-Lex. (2020). 52020PC0593—Proposal for a regulation on crypto-assets (MiCA). https://eur-lex.europa.eu/legal-content/EN/TXT/?uri=CELEX%3A52020PC0593

Federal Government of Germany. (2023). *Key impetus for new growth.* https://www.bundesregierung.de/breg-de/service/archiv-bundesregierung/financing-for-the-future-act-2247606

Federal Gazette, Federal Republic of Germany. (2024). Gesetz über die Digitalisierung des Finanzmarktes (FinmadiG), Bundesgesetzblatt I 2024 Nr. 438, 27 December 2024.

Federal Ministry of Finance of Germany. (2024). *FinTech Made in Germany.* https://www.bundesfinanzministerium.de/Content/EN/Standardartikel/Press_Room/Publications/Brochures/fintech-made-in-germany.pdf

Ferretti, F. (2022). Open banking: Gordian legal knots in the uncomfortable co-habitation between the PSD2 and the GDPR. *European Review of Private Law, 30*(1), 73-102. https://doi.org/10.54648/erpl2022004

Finastra. (2023). *Financial institutions in Germany see Gen AI and BaaS as next big disruptors to financial services, says Finastra survey.* https://www.finastra.com/press-media/financial-institutions-germany-see-gen-ai-and-baas-next-big-disruptors-financial

FinTech İstanbul. (2025, April 12). *Güncel verileri ile Türkiye'de "dikeylerine göre" FinTech şirketleri.* https://fintechistanbul.org/2025/04/12/guncel-verileri-ile-turkiyede-dikeylerine-gore-fintech-sirketleri/

Geisler, M., Mascarenhas, M. C. D., & You, Q.-Y. (2025). *Fintech 2023/2024.* BDO. https://www.bdo.de/en-gb/insights/updates/branchencenter/fintech-2023-2024

German Banking Industry Committee. (2025). GBIC comments on trilogue negotiations on the EU Payment Services Regulation (PSR) and PSD3. Bankenverband. https://bankenverband.de/en/zahlungsverkehr/gbic-comments-trilogue-negotiations-eu-payment-services-regulatio

Germany Trade & Invest. (2025). FinTech in Germany.

Germany Trade & Invest. VC investors rediscover German fintech. https://www.gtai.de/en/invest/industries/vc-investors-rediscover-german-fintech--1908118

Gounari, M., Stergiopoulos, G., Pipyros, K., et al. (2024). Harmonizing open banking in the European Union: An analysis of PSD2 compliance and interrelation with cybersecurity frameworks and standards. *International Cybersecurity Law Review, 5*, 79–120. https://doi.org/10.1365/s43439-023-00108-8

Gürkaynak, G., Yukaruç, N. N., & Eroğlu, N. A. (2021). *Regulation on sharing of secret information.* Lexology.

Hanten, M., Nalbantoglu, L., Polat, Y., & Tumer, B. (2022, June 28). *Digital banking and banking as a service—A comparison of Turkish and European law.* SSRN. https://ssrn.com/abstract=4148485

Herrera, L. A. (2025). PSD3 and the regulation on payment services in the context of crypto assets as a means of payment. In Pastor Sempere, C. (Ed.), *Governance and Control of Data and Digital Economy in the Eu-ropean Single Market: Vol 71. Law, Governance and Technology Series.* Springer. https://doi.org/10.1007/978-3-031-74889-9_8

Heuking Kühn Lüer Wojtek. (2025). *Future Financing Act II—Promotion of investments in renewable energies.* https://www.heuking.de/en/news-events/newsletter-articles/detail/zukunftsfinanzierungsgesetzes-ii-foerderung-von-investitionen-in-erneuerbare-energien.html

IHK Frankfurt am Main. (n.d.). *Länderzuständigkeiten §34c GewO.* https://www.frankfurt-main.ihk.de/branchenthemen/bau-und-immobilienwirtschaft/brancheninfos-beratungsangebote/laenderzustaendigkeiten-34c-gewo-5329332

KPMG. (2025a, March). *Pulse of Fintech H2/24: Investment decline continues.* https://kpmg.com/de/en/home/insights/2025/03/pulse-of-fintech-h2-24.html

KPMG. (2025b). *Top fintech trends for H1'25.* https://kpmg.com/xx/en/our-insights/risk-and-regulation/top-fintech-trends-in-h1-2025.html

Krimphove, D. (2019). Fintechs: Rechtliche Grundlagen moderner Finanztechnologien. (Chapter 35). Schäffer-Poeschel. https://doi.org/10.34157/9783791045979

Krishnakumar, V. (2025). *CloudOptimo. The True Cost of Cloud Data Egress and how to manage it.* https://www.cloudoptimo.com/blog/the-true-cost-of-cloud-data-egress-and-how-to-manage-it/

Kunz, J. H. (2022). *The financial technology law review* (5th ed.). Noerr. https://www.noerr.com/en/-/media/files/web/files-vor-rebranding-2023/news/2022/the-financial-technology-law-review---fifth-edition.pdf

Loonen, T., & Janssen, R. (2023). Implementation of MiFID II investor protection provisions by private banks within the European Union. *Journal of Financial Regulation and Compliance, 31*(1), 1–15. https://doi.org/10.1108/JFRC-10-2021-0087

Martínez Nadal, A. (2025). Stablecoins in the MiCA Regulation, In C. Pastor Semper e(Ed.), *Governance and control of data and digital economy in the European single market: Vol. 71, Law, Governance and Technology Series.* Springer. https://doi.org/10.1007/978-3-031-74889-9_8

Mkrtchyan, G., & Treiblmaier, H. (2025). Business implications and theoretical integration of the Markets in Crypto-Assets (MiCA) Regulation. *FinTech, 4*(2), 11. https://doi.org/10.3390/fintech4020011

Mock, S. (2020). Germany. In D. Busch, G. Ferrarini & J. P. Franx (Eds.), *Prospectus regulation and prospectus liability.* Oxford University Press. https://doi.org/10.1093/law/9780198846529.003.0020

Munich Start-up. (2021). *Munich is the second largest fintech location in Germany.* https://www.munich-startup.de/en/68474/munich-is-the-second-largest-fintech-location/

Ozili, P. K. (2025). Artificial Intelligence application and research in accounting, finance, economics, business, and management. In Chelliah, P. R., Dutta, P. K., Kumar, A., Gonzalez, E. D. R.

S., Mittal, M., & Gupta, S. (Eds.), *Generative Artificial Intelligence in Finance*. Scrivener Publishing. https://doi.org/10.1002/9781394271078

Onay, C., Öztürkkal, B., & Yıldırım, C. (2024). *Türkiye Fintech Research*. KPMG Türkiye & FINTR, PRTFO.

Ovenc, G., & Nabiyev, A. B. (2025). Discover how fintech is transforming bank performance: Insights from an emerging economy. *Cogent Economics & Finance, 13*(1), 2477676. https://doi.org/10.1080/23322039.2025.2477676

Parente, F. (2024, February). *The AI Act and its impacts on the European financial sector*. Eurofi Magazine. European Insurance and Occupational Pensions Authority. https://www.eiopa.europa.eu/system/files/2024-02/ai-act-fausto-eurofi-magazine_ghent_february-2024%20128.pdf

Perçin, Ö. (Ed.). (2023). Sorularla Fintek (2. baskı). On İki Levha Yayıncılık. ISBN 978-6254327209.

Polasik, M., Butor-Keler, A., Widawski, P., & Keler, G. (2024). Evaluating the regulatory approach to open banking in Europe: An empirical study. *Financial Law Review, 34*(2), 59–90. https://czasopisma.bg.ug.edu.pl/index.php/flr/article/view/11427

Presidency of the Republic of Türkiye Finance Office. (2024). The state of fintech ecosystem in Türkiye 2024. Presidency of the Republic of Türkiye Finance Office. https://www.cbfo.gov.tr/

Şahin, B. Ş., & Cantürk, B. C. (2020). Türkiye'deki hukuki altyapı ve Ödeme Hizmetleri Yönergesi 2 bağlamında API teknolojisi ve açık bankacılık [API technology and open banking in the context of the Payment Services Directive 2 in Türkiye]. *Maliye ve Finans Yazıları, 35*(114), 149–180. https://doi.org/10.33203/mfy.637661

Schroeder, H. (2008). Unfair Competition. In M. Wendler, B. Tremml & B. Buecker (Eds.), *Key Aspects of German Business Law*. Springer. https://doi.org/10.1007/978-3-540-68577-7_12

Sempere Pastor, C. (2025). Crypto assets and financial data space regulation in the EU's hybrid system of hard and soft law. In Pastor Sempere, C. (Ed.), *Governance and Control of Data and Digital Economy in the European Single Market: Vol 71, Law, Governance and Technology Series*. Springer. https://doi.org/10.1007/978-3-031-74889-9_8

Škrabka, J. (2024). Modernising payment services and enhancing open banking: A comparison of recent EU proposals of Payment Services Directive 3 (PSD3) and Payment Services Regulation (PSR) with current PSD2. Horizons of Law in Business and Finance, pp. 217–226.

Sönmez, Günaydin, Galatali, Kalinoglu. (2024). *Erdem & Erdem. Legal Landscape of FinTech in Türkiye: Trends and Transformations in 2022 & 2023*. https://www.erdem-erdem.av.tr/en/insights/legal-landscape-of-fintech-in-turkiye-trends-and-transformations-in-2022-and-2023

Taylor Wessing. (2024). 2025 *FinTech Outlook: 5 key areas to watch*. https://www.taylorwessing.com/de/insights-and-events/insights/2024/02/2024-fintech-outlook#:~:text=In%202025%2C%20fintech%20companies%20specializing,on%20the%20use%20of%20stablecoins

Tomasz, T. (2022, July). Crypto-assets and crypto-assets' subcategories under MiCA Regulation. *Capital Markets Law Journal, 17*(3), 365–382, https://doi.org/10.1093/cmlj/kmac008

TUIK. (2025, October 25). *Median age by country 2025*. World Population Review. https://worldpopulationreview.com/country-rankings/median-age

Westfall, R., & Dickens, S. (2025). *HyperFrame Research. Why are Cloud Egress Fees Really a Thing?* https://hyperframeresearch.com/2025/09/12/why-are-cloud-egress-fees-really-a-thing/

Yıldırım, S. (2024). Bankacılıkta uzaktan kimlik tespitinde karşılaşılan riskler ve çözüm önerileri. *Mülkiye Dergisi, 48*(1), 243–276.

Zetzsche, D., & Woxholth, J. (2025). Scope of MiCA. In *The EU law on crypto-assets: A guide to European FinTech regulation* (pp. 40–60). Cambridge University Press.

The Role of New Fintech Solutions and Related Regulations in Improving Access to Finance for Consumers and SMEs

Iclal Nihal Barac Evci

1 Introduction

Small and medium-sized enterprises (SMEs) are the economic core of most developed and developing countries (World Bank Group, 2022). Globally, SMEs account for over 90 percent of businesses and over 50 percent of employment. They are also substantial contributors to economic activity, typically accounting for 60 to 70 percent of a country's GDP (World Bank Group, 2022).

However, many factors prevent SMEs from accessing adequate finance. These include the higher cost of reaching and serving SMEs relative to the revenue potential for traditional banks; information asymmetries that result in the unavailability of financial and credit data needed by lenders to assess the creditworthiness of SMEs; and, on the SMEs' side, lack of collateral, lack of financial literacy, lack of physical access to bank branches, and difficulties in registration and verification (Bazarbash & Beaton, 2020; World Bank Group, 2022). Approximately half of formal SMEs lack access to formal credit, depending instead on internal funds or cash from friends and family to start and initially operate their businesses. A comprehensive survey of SMEs across 135 countries found that limited access to financing was the most significant obstacle to their continued operations. In emerging markets, around 131 million SMEs, accounting for 41% of the total, have unmet financing needs (Bazarbash & Beaton, 2020; World Bank Group, 2022). Women-owned businesses account for 23% of SMEs and 32% of the SME financing gap (World Bank Group, 2022).

Barriers to accessing credit are also a prevalent issue among consumers. According to the World Bank, around 60 percent of adults in developing countries do not use conventional financial services. Creating opportunities for SMEs and consumers in

I. N. Barac Evci (✉)
Faculty of Law, Department of Commercial Law, Piri Reis University, Istanbul, Türkiye
e-mail: nbarac@pirireis.edu.tr

Future Finance Law Hub, Istanbul, Türkiye

B. Ş. Şahin and B. C. Cantürk (eds.), *Fintech and Sustainable Development*, Sustainable Finance, https://doi.org/10.1007/978-3-032-31020-0_5

emerging markets will not only promote economic development and reduce poverty, but will also indirectly support women's labor-market participation (Bazarbash & Beaton, 2020).

The World Bank Group (2022) asserts that FinTechs can help reduce the financing gap for SMEs and consumers by providing access to alternative funding sources and enhancing access to traditional financial institutions through the development of innovative digital products and process automation. Fintech solutions encompass mobile payment applications, digital banking platforms, and equity products such as supply chain finance, secured revolving lines of credit, merchant receivables financing, peer-to-peer (P2P)/marketplace lending, and crowdfunding (World Bank Group, 2022). Furthermore, the digital transformation of internal and business-to-business (B2B) processes, such as electronic invoicing (e-invoicing) and tokenized assets using blockchain/distributed ledger technology (DLT), can also help mitigate barriers to SME access to finance (World Bank Group, 2022).

Fintechs can also facilitate consumers' access to finance. In particular, mobile payment apps, digital bank loans, and P2P lending systems offer consumers unprecedented convenience, efficiency, and accessibility in managing their finances (Iyelolu et al., 2024; Miriyeva, 2022). However, a counterargument is that the rise of fintech solutions and mobile payment methods has reduced cash use. At the same time, the transition to a cashless society has hindered the homeless population's access to financial services, as many lack Internet access and mobile devices (Lupo-Pasini, 2021).

All of these rapid technological innovations and emerging business models in the FinTech sector have rendered the existing regulatory framework inadequate or created regulatory gaps. Moreover, existing regulatory barriers and compliance requirements can create significant obstacles for FinTech start-ups, hindering their ability to innovate and compete effectively with incumbent financial institutions (Iyelolu et al., 2024). The high initial capital requirements in the banking sector, detailed reporting requirements, and the redundancy and complexity of secondary regulations issued by regulatory authorities are unfavorable and create barriers to entry for fintechs, which are often founded by a small number of technologists with limited capital (Iyelolu et al., 2024). On the other hand, current regulations may not adequately address the unique risks and characteristics associated with FinTech activities, such as data privacy concerns, cybersecurity threats, and algorithmic biases (Iyelolu et al., 2024; World Bank Group, 2022).

In response to these challenges, regulatory authorities have increasingly adopted a more flexible and adaptive approach to FinTech regulation. One of the most significant benefits of FinTech regulatory reform is its role in promoting greater financial inclusion by expanding access to banking and credit services, particularly for underserved populations (Iyelolu et al., 2024). By reducing regulatory barriers and promoting competition and innovation in the financial services sector, this reform should enable millions of unbanked and underbanked individuals to participate in the formal financial system, thereby promoting greater financial inclusion and socioeconomic empowerment (Iyelolu et al., 2024).

Currently, it is observed that certain innovative fintech solutions are at the forefront of supporting both SMEs and consumers. Many jurisdictions prefer to adopt regulations in the context of these new products and services. It is understandable that regulations based on products and services facilitate the observation of specific products and services, their effects on reality, their development points, and areas of misuse. Lawmakers are therefore able to specify the areas to be regulated effectively. For these reasons, the following section of the study analyzes the leading fintech solutions that effectively enable SMEs and consumers to access finance, as well as the comparative regulatory regimes associated with them. This section is primarily based on a comparative analysis of Turkish and European law, but also includes illustrative examples of legal regulations from various countries, which are noteworthy. Within this scope, the second section includes the comparative regulations concerning digital payments, digital banks, open banking, and equity crowdfunding in sequence.

2 Fintech, Regulatory Framework and Financial Access

As we mentioned in the introduction, the fintech solutions supporting SMEs and consumers have been concentrated on specific products and services. Our research and observations have indicated that the most prominent products and solutions are digital payments, digital banking, open banking, and equity crowdfunding at present. In the following part of this section, each type of solution will be defined, and the important regulatory information regarding these solutions will be explained.

2.1 Digital Payments

Digital payments refer to the transfer of value facilitated by digital or electronic devices and channels for transmitting data. Digital payments encompass payments initiated via a debit or credit card, a mobile phone, a computer, a tablet, or a wearable digital device (World Bank Group, 2022). *Mobile payments,* as a subset of digital payments, represent one of the most prominent sectors within the FinTech ecosystem. Innovations affecting digital payment technologies, such as mobile wallets, electronic money, and contactless payments, have revolutionized the way individuals and businesses conduct transactions.

Electronic money (e-money) has emerged as a pivotal instrument in the digital payments landscape. Electronic money is a digital alternative to cash that enables the storage of funds on a device (e.g., a card or a mobile phone) or via the Internet, as well as the execution of payment transactions (Perçin et al., 2023a; Pouelle et al., 2024; Sünbül, 2023).

One of the most important services required for the effective use of electronic money is *digital wallets.* A digital wallet is a payment tool offered as an electronic device, online service, or application that stores information about the payment account or payment method defined by the customer and allows the customer to make payments using this information. In other words, customers can save their existing account or card information to their digital wallets and use these accounts and cards without needing to log into their bank's online banking or mobile application (Sünbül, 2023). The emergence of digital wallet services provided by e-money institutions has profoundly transformed the landscape of financial transactions. These services enable individuals to store electronic money in digital wallets, facilitating cashless payments, money transfers to and from various accounts, and a broad range of financial activities. The use of digital ID and alternative identification methods enables secure access for individuals without bank accounts, allowing them to store their financial assets in digital accounts (Miglionico, 2022).

These technological advancements have enabled individuals, particularly those in rural areas, to conduct financial transactions without visiting bank branches or ATMs (Wamba-Taguimdje & Kamdjoug, 2024). Additionally, the use of technology to facilitate the swift transfer of remittances at minimal cost (including, as a key innovation, foreign exchange (FX) conversion) in a secure environment can have a substantial impact on financial inclusion, particularly in the context of remittances sent by migrant workers to support their families in their home countries, which often face high transfer fees (Miglionico, 2022).

Digital payments have the potential to offer significant benefits to SMEs by reducing transaction and delivery costs for commercial customers, as well as fraud risks and potential chargebacks (Wamba-Taguimdje & Kamdjoug, 2024; World Bank Group, 2022). Advancements in technology, including mobile phones and the Internet, contactless methods such as QR codes, Big Data analytics, APIs, and biometric technologies, are contributing to the evolution of digital payments. These innovations have enabled the emergence of new delivery channels and payment methods that significantly reduce the complexity of the payment process, facilitate customer onboarding, and enhance the efficiency of real-time approvals (World Bank Group, 2022). To provide an example from current practice, e-money institutions operating as PF (Payment Facilitator) can reduce transaction costs for SMEs by automatically determining the lowest commission rates offered by banks or e-money institutions that have customer acquisition and card acceptance licenses, and processing transactions through this institution when intermediating payment transactions of member merchants via virtual POS.

Furthermore, transaction data from digital payments can provide valuable insights into SME credit risk, thereby helping mitigate the information asymmetry barrier (Huang et al, 2020). Digital transaction data can be used by fintech firms, banks, and large technology companies to assess merchants' credit risk. This assessment can then be used to determine whether to offer loans and other financial products to small businesses that have previously lacked access to financing. This is particularly beneficial in developing economies, where many small businesses lack formal financial information or have previously lacked access to traditional financial products (Huang et al, 2020; Miglionico, 2022; World Bank Group, 2022).

On the other hand, potential fraud and theft should be considered by SMEs when adopting digital payments. Given the rapid nature of digital payments, fraudsters may be able to withdraw funds before their fraudulent activity is identified. Furthermore, the World Bank Group (2022) has identified the growing use of new devices, including mobile phones and IoT devices, as a potential vulnerability that hackers could exploit.

The vast majority of these topics examined in the literature are regulated by law in both Turkish and EU law. In Türkiye, mobile payments are regulated by the Code No. 6493 on Payment and Securities Settlement Systems, Payment Services, and Electronic Money Institutions, which shall henceforth be referred to as the Turkish E-Money Code. The Regulation on Payment Services and Electronic Money Issuance and Payment Service Providers, which shall henceforth be referred to as the Turkish E-Money Regulation, was prepared in accordance with this law. In addition, there is also the Notice on Information Systems of Payment and Electronic Money Institutions and Data Sharing Services in the Field of Payment Services of Payment Service Providers dated December 1, 2021.

In European Union law, payment services are currently regulated by the Payment Services Directive (EU) 2015/2366, known as PSD2. PSD2 is an attempt to decentralize and regulate the payments industry, to give space to small- and medium-sized companies while ensuring consumer transparency and further protection (Arıcı, 2020; Bispo, 2020; Çeşit, 2021). There is also Directive 2009/110/EC of the European Parliament and of the Council on the Taking up, Pursuit and Prudential Supervision of the Business of Electronic Money Institutions (hereinafter referred to as the EU E-Money Directive), which regulates aspects specific to electronic money institutions. However, the Commission announced on June 28, 2023, that it would modernize both Directives by merging them into a single piece of legislation and harmonizing them to the extent possible (European Commission, 2023b; Pouelle et al., 2024).

Under Turkish law, pursuant to Article 13 of the Turkish E-Money Code, banks, electronic money institutions, payment institutions, and the Post and Telegraph Organisation Joint Stock Company (PTT A.Ş.) are deemed as *payment service providers*. Other persons cannot provide payment services. This situation is quite similar in EU law, where credit institutions, payment institutions, and e-money institutions are also considered payment service providers (Pouelle et al., 2024). Among the mentioned institutions, *electronic money institutions (EMIs)* and *payment institutions (PIs)* are important to this report, as they can be considered fintech institutions. These institutions are defined quite similarly in EU and Turkish legislation.

An electronic money institution is defined as a legal person authorized to issue electronic money (Turkish E-Money Code, Art. 3/1-k; EU E-Money Directive, Art. 2/1). A payment institution is defined as a legal person that has been authorized under the Code to provide and execute payment services (Turkish E-Money Code, Art. 3/1-oo; PSD2, Art. 4/4; Pouelle et al., 2024). Electronic money is defined as an electronically stored monetary value as represented by a claim on the issuer, issued upon receipt of funds for the purpose of making payment transactions as defined in law, and accepted as a payment instrument by a natural or legal person other than the electronic money issuer (Turkish E-Money Code, Art. 3/1-i; EU E-Money Directive,

Art. 2/2). In simpler terms, e-money institutions enable electronic money to be held in an electronic account. In contrast, payment institutions enable payments by offering payment services specified in the Turkish E-Money Code, Art. 12 (Perçin et al., 2023a). Under Turkish law, these two services require separate licenses (Sünbül, 2023). However, since these services make sense together, it has been observed that fintech companies in Türkiye generally operate both services. Since EU law already has two separate Directives for these two institutions, the same applies to them. However, PSD3 proposes a new PSP category called payment institutions providing electronic money services, thereby eliminating this dichotomy (Herrera, 2025).

In an e-money institution, the issued electronic money is technically stored in a digital account, and the customer makes payment transactions using the balance in this account (Akıncı, 2024; Perçin, et al., 2023a). In addition to digital accounts, payment and electronic money institutions may also offer digital wallet services (Perçin et al., 2023a). The digital wallet service to be provided by the institutions is defined in detail in Article 4A of the Turkish E-Money Regulation. According to this Regulation, digital wallet services may only be provided by payment service providers. Only the payment account with the payment service provider offering the service, and the electronic device, online service, or application where the payment instrument issued by this payment service provider can be identified and stored, are not considered digital wallet services. For a digital wallet to be used as a payment instrument and for funds related to payment transactions to be transferred through an institution providing digital wallet services, the institution must also have the authority to issue electronic money. For a digital wallet to be used directly with another payment service provider's payment account or for a payment instrument issued by another payment service provider to be used in a payment transaction, the relevant institution must have the authority to provide the "payment initiation service," which is examined separately under the heading "Open Banking" below (Perçin et al., 2023a).

Under EU legislation, e-wallet providers are currently considered technical service providers and are therefore out of scope of the PSD2 (PSD2, Art. 3-j; Divissenko & D'Amico, 2025). Conversely, unlike other back-end technical services, e-wallets have now become the key interface through which consumers manage their online banking in the EU as well. Technical service providers manage payment transaction data, support strong customer authentication, and, in the future, determine how information on charges is communicated to consumers. However, technical service providers are not within the scope of the Payment Services Regulation, except for being liable if they fail to support strong customer authentication. This means that technical service providers do not have any obligations on how to provide information to consumers (e.g., information on charges for money remittances) (BEUC, 2023). One of the objectives of PSD3 is to enhance the efficacy of digital payment wallets, which are electronic wallets that store virtual payment cards, by mandating the implementation of SCA (Strong Customer Authentication) during the enrollment of a payment instrument in the wallet. This responsibility belongs to the PSPs (Payment Service Providers) that issued the instrument (European Commission, 2023b; Sünbül, 2023).

Another important point about the e-money and payment institutions is that these institutions are obliged to transfer the funds collected in exchange for the issuance of electronic money in a specific account defined as a "safeguarding account" to be opened within the banks (EU E-Money Directive, Art. 7; PSD2, Art. 10; Pouelle et al., 2024). In practice, this obligation has created several challenges for e-money and payment institutions. For instance, commercial banks sometimes refuse to open bank accounts for these parties or close existing accounts due to concerns about anti-money laundering controls (Çeşit, 2021). In order to overcome this difficulty, the requirements for banks regarding bank account services for non-bank payment service providers will be significantly tightened under PSD3, with a more substantial obligation for banks to explain any denial of access, including the withdrawal of service (European Commission, 2023b; Herrera, 2025). The basis for denying the request must be derived from the particular circumstances pertaining to the PI in question, taking into account any substantial evidence indicating the potential engagement in illicit activities by or through the PI, or a business model or risk profile that poses a significant threat to the credit institution. The PI will have the right to submit an appeal to a national authority against any decision by a financial institution to deny or cancel an account. In addition to commercial banks, central banks may also offer account services to non-bank payment service providers (PSPs) at their discretion (European Commission, 2023b). Under Turkish law, the opening of safeguarding bank accounts is regulated in detail in Articles 34, 35, and 36 of the Regulation on Payment Services and Electronic Currency Issuance and Payment Service Providers. However, the existing regulations address issues such as the deposit of funds with the bank, reconciliation, and banks' authority to block funds to protect the rights of fund holders, but do not address banks' obligation to open accounts (Çeşit, 2021). However, in practice, e-money institutions have not encountered any difficulties in this respect.

A similar challenge faced by e-money and payment institutions is the need to access key payment infrastructure to execute and settle payments in order to offer payment services. In the EU, Directive 98/26/EC on Settlement Finality in Payment and Securities Settlement Systems, in its current form, prevents PIs and EMIs from gaining access to the payment infrastructures identified under the Directive by failing to specify them as possible participants in such systems. This dynamic has led to a situation where PIs and EMIs are increasingly reliant on commercial banks, creating a structural dependency and an uneven competitive landscape (Çeşit, 2021; Christensen, 2025). As banks are also competitors of PIs and EMIs in the provision of payment services, this relationship is complex and requires careful management (Christensen, 2025; European Commission, 2023b). The PSD3 proposal aims to resolve the above-mentioned issue by including PIs as potential participants in the designated payment systems. In addition, rules governing PI participation in payment systems will be strengthened, and payment system operators will be required to conduct appropriate risk assessments (European Commission, 2023b; Herrera, 2025).

Meanwhile, the European Parliament and Council's Directive (EU) 2024/886 of March 13, 2024, on Instant Credit Transfers in Euros, has made significant changes to

Directives 98/26/EC and (EU) 2015/2366. Under Article 5a(4)(c) of the EU Instant Credit Transfer Directive, PIs and EMIs are now recognized as institutions under Article 2 of Directive 98/26/EC, thereby enabling them to participate directly in the SEPA instant payment system, another EU legal innovation that allows for money transfers within 10 seconds. To comply with these regulations, a transition period has been granted until April 9, 2027, for PIs and EMIs within the Eurozone, and until July 9, 2027, for those outside the Eurozone (EU Instant Credit Transfer Directive, Art. 5a(8)). Therefore, some of the changes intended by PSD3 were made by Directive (EU) 2024/886.

Similarly, PIs and EMIs should carry out their activities through banks under the Turkish Law (Turkish E-Money Code, Art. 18/4; Art. 20/3). However, in Türkiye, as a similar significant development, payment and electronic money institutions have been granted the opportunity to participate in the Continuous and Instant Transfer of Funds (FAST), a payment system operated by the Central Bank of the Republic of Türkiye (TCMB, 2022). Thus, it can be claimed that these institutions' dependence on commercial banks for their operations has already reduced.

The final point we need to address regarding financial inclusion and EMIs and PIs is that Turkish law prohibits these institutions from engaging in credit activities (Turkish E-Money Code, Art 14/4, Art. 20/4; Sünbül, 2023). In addition, the electronic money issuer may not pay interest or provide any benefit to the electronic money holder depending on the period of holding the electronic money (Turkish E-Money Code, Art. 20/5). In EU law, the same rule applies, except that e-money institutions may only grant credit limited to the payment services activities they provide (EU E-Money Directive, Art. 6/1-b). However, there are specific conditions that must be met for this to be possible (Pouelle et al., 2024). These conditions are regulated in detail by Article 6(1)(b) of the EU Electronic Money Directive. Firstly, the credit must be supplementary and granted exclusively in association with the execution of a payment transaction. Secondly, the credit must be repaid within a brief period, not exceeding twelve months. Thirdly, the credit cannot be granted from the funds received or held for the purpose of executing a payment transaction. Lastly, the payment institution's own funds must be adequate and sufficient to cover the total amount of credit granted.

Currently, this is the most important difference between the EU law and the Turkish law on payment and e-money institutions. The exceptional credit services regulation in EU law allows consumers to buy a product and pay for it in instalments or defer the payment (known as the buy now pay later system), while SMEs can similarly pay in instalments or obtain credit in connection with a payment acceptance transaction. For example, fintechs such as Klarna[1] and PayPal Credit,[2] which are also active in Europe, allow consumers to pay for a product in instalments. For services that go beyond this scope, e-money and payment organizations also must operate as financial companies. For example, Stripe,[3] a fintech that offers a wide

[1] https://www.klarna.com/de/.

[2] https://www.paypal.com/us/digital-wallet/ways-to-pay/credit-services/paypal-credit.

[3] https://stripe.com/en-de/enterprise.

range of payment acceptance services (card, POS, digital wallet, or online shopping) to SMEs, provides small loans to businesses to be repaid on the payments received. Since the services offered by fintechs specifically to SMEs are very diverse (payment acceptance, money storage, financial automation, accounting and tax automation, payment tracking, etc.), lending services in this area are also becoming quite diverse.

The fact that e-money institutions and payment institutions in Türkiye are not authorized to provide any services that would constitute lending may hinder innovation in this area. In the current situation, fintechs are trying to provide such services in cooperation with financial companies or banks, or they are operating in this field by obtaining a digital banking license directly. For example, Paycell, a payment and e-money institution, offers BNPL services through Turkcell Finansman A.Ş.,[4] which is a group company. TURK Elektronik Para A.Ş., another e-money and payment service institution, offers BNPL and shopping credit services for shopping at contracted merchants through the financing company TURK Finansman A.Ş.,[5] which is a group company, and at the same time provides installment payment for payments made with credit cards of contracted banks in the payment acceptance service it provides through POS. These services provided by PIs and EMIs are in the nature of intermediating the payment of the credit received by the consumer from the financing company. These examples show that only fintechs that reach a certain size can provide credit services. Fintechs that cannot offer lending services other than these cooperation opportunities in Türkiye may also prefer to establish digital banking in Türkiye, which we will examine below.

2.2 *Digital Banks*

A digital bank is characterized as a credit institution that offers banking services through electronic distribution channels, instead of conventional physical branches (Gürses, 2025). Digital banks should not be confused with digital banking applications that enable banking services to be provided via the Internet using electronic devices such as phones and computers (Perçin et al., 2023b).

Digital lending platforms offer certain advantages over traditional banking institutions, such as faster application, review, and approval processes conducted online and through mobile devices (World Bank Group, 2022). For example, an empirical study of the Paycheck Protection Program (PPP), which was created by the U.S. government and offers guaranteed and potentially forgivable small business loans during the COVID-19 crisis, has shown that FinTech lenders often serve sectors and neighborhoods with less access to traditional financing. Neighborhoods with fewer bank branches, relatively lower median incomes, and larger minority populations receive a larger share of their loans from these new lenders (Erel & Liebersohn, 2022).

---

[4] https://financell.com.tr/.

[5] https://param.com.tr/hakkimizda.

Despite the increasing role of FinTech in the financial services industry, allowing financial intermediation by FinTech lenders remains controversial (Erel & Liebersohn, 2022). Despite their capacity to operate at a higher speed than financial institutions, only a small number of the most prominent digital lending platforms are able to function in an active capacity due to the restrictive financial regulations that are in place in numerous countries. For instance, during the course of the pandemic, Chinese digital lending platforms collaborated extensively with banking institutions and other financial entities to distribute relief schemes. A notable example is the collaboration between MYBank and Alipay with over 100 local Chinese banks, which aims to provide small street vendors in Wuhan with rapid digital loans. MYBank has already disbursed loans to more than 350,000 merchants and is waiving interest rates for the first month (World Bank Group, 2022).

In Turkish law, digital banks are regulated by the regulation on the Operating Principles of Digital Banks and Service Model Banking, which shall henceforth be referred to as the Turkish Digital Banks Regulation. According to Art. 5/1 of the Turkish Digital Banks Regulation, the credit customers of digital banks can only be financial consumers and SMEs. Some criteria have been introduced for digital banks to obtain a license to establish and operate, one of which is that they must include a business plan that includes the needs identified for the target groups such as students, housewives, young people under the age of eighteen, SMEs, and the needs identified for the groups included in these target groups, the products and services they plan to offer to meet these needs, and their marketing strategy, which they have determined to increase financial inclusion (Turkish Digital Banks Regulation, Art. 11/1-a-1; Gürses, 2025). However, the total amount of unsecured cash consumer loans that can be granted by digital banks cannot exceed four times the average monthly net income of the respective customer declared and confirmed by the digital banks, and cannot exceed 10.000 Turkish Liras if the average monthly net income of the customer cannot be determined (Turkish Digital Banks Regulation, Art. 5/4). As can be seen, the development of inclusive solutions for SMEs and consumer groups with limited access to finance is explicitly regulated in Turkish law as one of the main objectives of digital banking systems, while on the other hand, it is aimed to protect the financial stability of digital banks by limiting the granting of large amounts of credit to these individuals.

According to Article 9 of the Turkish Digital Banks Regulation, the minimum paid-in capital required for digital banks to obtain an operating license is one billion Turkish Liras. According to Article 7/1-f of the Banking Law No. 5411, the minimum paid-in capital required to establish a traditional bank is 30 million Turkish Liras (Gürses, 2025). As can be seen, in the current situation, much more capital is required to establish a digital bank than to establish a traditional bank. However, since the cost of establishing and organizing branches in traditional banking requires regular expenditures, the establishment of a digital bank without the obligation to open a branch is still more advantageous since it does not require any expenditures.

Unlike in Türkiye, there is no specific legislation in the EU for digital banks, also known as neobanks. Therefore, institutions wishing to operate as digital banks usually start their operations by obtaining a standard banking license under Directive 2013/

36/EU of the European Parliament and of the Council of June 26, 2013, on access to the activity of credit institutions and the prudential supervision of credit institutions and investment firms. This is because, according to the provisions of Art. 33 et seq. of this Directive, a credit institution operating with a license obtained from a member state has the freedom to open branches or provide services in other member states (Pouelle et al., 2024; Cardinal & Thomas, 2022).

One of the most prominent examples in this context is N26,[6] which was founded in Germany. Initially established as a fintech, the company obtained a banking licence from the Federal Financial Supervisory Authority (Bundesanstalt für Finanzdienstleistungsaufsicht, BaFin) and, after reaching a certain customer base, continued its activities as a digital bank. Similarly, the Netherlands-based Bunq B.V.,[7] a fintech, obtained a banking license and operates as a digital bank. UK-based Revolut[8] is one of the most well-known examples of a digital bank that entered the EU market with a banking license obtained from Lithuania. Although there is no legal obligation, unlike the regulations in Türkiye, the target audience of digital banks in Europe is SMEs and consumers as well.

Although a specific regulation for digital banks has been issued in Türkiye, these banks are subject to the Banking Law No. 5411 in all matters other than those regulated in the regulation (Perçin, 2023b). Therefore, there is no major practical difference with the situation in EU law. In both jurisdictions, digital banks are subject to all the obligations imposed on traditional banks. In particular, provisions regarding corporate governance and detailed reporting obligations are seen as the main challenges for the operation of digital banks.

2.3 *Open Banking*

Open banking refers to the opening up of banking products and services to a wider audience and the development of innovative financial products and services by making customer data held by banks and other financial institutions available to third parties for access and use at the customer's request (Jeng, 2022). Customer consent is required for data sharing, which is essential in this system. Open banking, on the one hand, constitutes an exception to the obligation to keep customer secrets confidential, but on the other hand, it facilitates the emergence of personalized products and services and the provision of cheaper, new, effective, competitive, and innovative services to customers, thereby increasing financial inclusion and contributing to the construction of a new financial services ecosystem (Casolaro et al., 2025; Gürses, 2025; Perçin et al., 2023b).

Open banking enables banks to provide basic banking services such as account viewing, initiating payment orders, and credit usage through third parties, allowing

[6] https://n26.com/en-eu.
[7] https://www.bunq.com/.
[8] https://www.revolut.com/de-DE/.

individuals to access banking services from different platforms, such as social media platforms and e-commerce sites that do not belong to the bank (Karakoç, 2024). This phenomenon is currently being pursued by at least 80 countries worldwide, with variations in implementation timing, data scope, and framework (Casolaro et al., 2025). Today, one of the most important business models in open banking is Banking as a Service (BaaS). In this system, banks offer their infrastructure and licenses to fintech companies via APIs, enabling fintech companies to use the BaaS infrastructure to offer banking services such as account opening, payment services, card issuance, and IBAN provision to their customers on their own behalf (Perçin et al., 2023b).

From a theoretical perspective, open banking applications are not a fintech solution, but they are one of the most important factors facilitating the activities of fintechs. This is because fintechs' activities in the field of payment systems, which is one of their most important areas of activity, are effectively made possible through open banking applications by means of access to bank accounts (Karakoç, 2024).

The most commonly used method in practice for conducting open banking activities is API (Application Programming Interface) sharing. An API, in its simplest definition, is an application programming interface created to enable one software program to use functions defined in another software program. Functions or data can be borrowed from one software program in real time using APIs, enabling interaction between different software programs (Cardinal & Thomas, 2022; Karakoç, 2024). To illustrate this with an example, when a user wants to make a purchase on an e-commerce site, after entering their card details on the payment screen, the payment process is completed in the background through APIs operating between the e-commerce site and the payment service provider (Bispo, 2020; Karakoç, 2024).

While the use of APIs is quite beneficial in terms of completing transactions quickly, there are also certain risks involved, such as unauthorized monitoring of the connection between the API client and the server, IP address abuse, and infrastructure failures (Wolberg-Stok, 2022). To prevent all these risks, it is important to take security measures such as end-to-end encryption, authentication, vulnerability testing, and an IP whitelist that only allows specific IP addresses to access the system (Karakoç, 2024).

In the Turkish and EU systems, it is required that the party holding the data share it upon request from an institution authorized by the customer to provide open banking services. This aims to enable existing products and services to be offered at lower costs and higher quality, to improve new products and services, and to strengthen competition in the sector. This aims to prevent the monopolization created by traditional banks offering all financial services together and to increase the quality of products and services by supporting fintechs, which generally offer one or a few financial products (Casolaro et al., 2025; Karakoç, 2024).

The fundamental concepts related to open banking in EU law are outlined in PSD2, which we have previously referred to in this section. With PSD2, the definitions of Account Information Service Providers (AISP) and Payment Initiation Service Providers (PISP), which are fundamental elements of open banking, have been established (Arner et al., 2022). According to Article 4 of the PSD2, the term "account information service" refers to an online service that provides consolidated information regarding one or more payment accounts held by the payment service user, either with another payment service provider or with multiple payment service providers. On the other hand, payment initiation services refer to the processes that facilitate the initiation of a payment order on behalf of a payment service user with respect to a payment account held at a different payment service provider. The provision of account information and payment initiation services is exclusively authorized to institutions that meet specific criteria (Arner et al., 2022). Another important point that was regulated in PSD2 is that banks are required to share the data of consenting customers with non-bank service providers (Perçin et al., 2023b). This information sharing serves two primary purposes: the analysis of consumer financial data and the facilitation of payment initiation accounts (Casolaro et al., 2025). Clearly, AISPs and PISPs need this information in order to perform their activities. It is also believed that introducing standardized technical solutions through such regulations will provide protection against misuse by parties in open banking (Arner et al., 2022; Bispo, 2020; Zunzunegui, 2018).

According to Article 66(1) of PSD2, Member States are required to ensure that payers are entitled to use a payment initiation service provider in order to access the payment services listed in point (7) of Annex I. Similarly, under Article 67(1), Member States must guarantee that payment service users have the right to use providers offering account information services as described in point (8) of Annex I. Furthermore, pursuant to Article 94(2), payment service providers may access, process, and store only the personal data that is strictly necessary for delivering their payment services, and they may do so only on the basis of the explicit consent of the payment service user. The obligation of banks to share customer data is based on this last provision. While it is not explicit in the PSD2, the Euro Retail Payments Board (ERPB) is advising the use of those APIs that are to be regarded as the best tool to comply with the requirements of the legislation (Bispo, 2020; Cardinal & Thomas, 2022). Additionally, Article 97 of the Directive requires payment service providers to implement strong customer authentication when customers access their payment account online, initiate an electronic payment transaction, or perform any action that may involve the risk of payment fraud or other abuse through a remote channel. This provision is also important in terms of the security of open banking practices (Bispo, 2020).

In Turkish law, the first regulation addressing open banking is Article 3 of the "*Regulation on Banks' Information Systems and Electronic Banking Services*" dated 2020. According to this provision, "*Open banking services: Refers to the electronic distribution channel through which customers or parties acting on behalf of customers can remotely access the financial services offered by the bank using methods such as API, web services, or file transfer protocols to perform banking transactions or give instructions to the bank to perform such transactions.*" Similarly, Article 41 of the same Regulation contains a provision regarding identity verification and transaction security in open banking services, which states that: "*When using open banking services, communication between the customer or the party acting on behalf of the customer and the bank must be end-to-end secure communication, and single-factor authentication shall not be considered a violation of the first paragraph of Article 34 (two-factor authentication), provided that the bank applies compensatory additional controls and imposes additional restrictions on the sources to which the customer can connect.*"

Furthermore, this regulation states that the BDDK (Bankacılık Düzenleme ve Denetleme Kurumu/ Banking Regulation and Supervision Board) is authorized to determine the procedures and principles related to open banking, and based on this authority, the BDDK issued the Regulation on the Operating Principles of Digital Banks and Service Model Banking (Turkish Digital Banks Regulation) in 2021. This regulation essentially focuses on digital banking, which we examined in a previous section of this study. However, the regulation contains detailed provisions regarding banking as a service (BaaS). As previously mentioned, in BaaS, banks provide fintech companies with infrastructure and licenses through APIs, enabling fintech companies to offer banking services such as opening accounts in their own name, providing payment services, card issuance, and offering IBANs to their customers (Perçin et al., 2023b).

In Article 3/1-jj of the Turkish Digital Banks Regulation, BaaS is defined as "*a service model whereby customers can connect directly to service banks' systems via open banking services provided through interfaces offered by interface providers and perform banking transactions through the service bank.*" According to that Regulation, in BaaS, the products and services that a bank can offer as an interface provider for a service bank are limited to the scope of its own operating license. On the other hand, to collect deposits or participation funds through an interface provider, this interface provider must be a payment or electronic money institution. This means that the interface provider may offer the customer the opportunity to deposit funds or participation funds with the service bank, but if the interface provider is not a payment or electronic money institution, the amount to be deposited must not enter the interface provider's accounting records in any way; it must be sent directly from the customer to the service bank or through institutions authorized to act as intermediaries for payment services (Karakoç, 2024; Perçin et al., 2023b). Therefore, it is difficult for fintechs that do not hold a payment institution or e-money institution license to provide effective services as interface providers. Indeed, in practice, interface providers are typically e-money and payment institutions. Banks cannot act as interface providers; they can only provide services as service banks. The regulation also assigns the responsibility of ensuring that the interface provider's electronic service channels and IT infrastructure meet adequate security criteria to the service bank. Provisions have been adopted stating that the interface provider and the service bank are jointly and severally liable for establishing controls within this scope. Furthermore, the responsibility for verifying the customer's identity also belongs to the service bank (Karakoç, 2024; Perçin et al., 2023b). To the best of our knowledge, there are no similar provisions in EU law regarding banking as a service, since there is currently no specific regulation addressing banking-as-a-service (BaaS) as such.

Finally, regarding the payment services aspect of open banking, in order to comply with the provisions on "Payment Initiation Service" and "Account Information Service" in the PSD2 regulation, the definition of payment services has been expanded to include open banking services through an amendment to Article 12 of Law No. 6493 on Payment And Securities Settlement Systems, Payment Services, And Electronic Money Institutions, the above Turkish E -Money Code. Paragraphs (f) and (g) of Article 12(1) of the Turkish E-Money Code define payment services as follows:

> *"f) Payment initiation services provided at the request of the payment service user in relation to a payment account held with another payment service provider,*
>
> *g) The service of providing consolidated information regarding one or more payment accounts held by the payment service user with payment service providers on online platforms, subject to the payment service user's consent."*

Subsequently, to demonstrate the implementation of these articles, the Regulation on Payment Services and Electronic Money Issuance and Payment Service Providers dated 01.12.2021 (referred to above as Turkish E-Money Regulation) has regulated the roles, responsibilities, and obligations related to these services, provisions regarding licensing, and the basic operational and technical rules for initiating payment orders and account information services (Perçin et al., 2023b).

2.4 Equity Crowdfunding

Equity crowdfunding can be defined as the process of raising modest amounts of equity capital from numerous retail investors ("the crowd") by issuing shares, generally via an online platform. The fundamental principle of crowdfunding is to facilitate the ability of small businesses to reach a large number of potential investors and thereby receive investment offers for their business (Kahveci, 2023; Korkmaz, 2020). Crowdfunding increases the promotional potential of offers through the use of technological tools, while reducing the costs of initial public offerings (IPO) (World Bank Group, 2022).

Equity investment through crowdfunding is an important tool for accessing capital, particularly for SMEs and start-ups (Korkmaz, 2020). The promotion of SMEs on online platforms has been demonstrated to enhance their visibility. Moreover, numerous regulatory authorities mandate that platforms undertake a certain degree of due diligence on SME issuers, thereby mitigating the information asymmetry between retail investors and SME issuers and empowering investors to make more informed investment decisions (World Bank Group, 2022).

Another important aspect of equity crowdfunding as a viable option for SMEs is the lower cost of regulatory compliance (World Bank Group, 2022). The prospectus issuance, authorization, reporting, and corporate governance requirements for issuing securities and going public under the traditional capital regime are costly for SMEs (Korkmaz, 2020; World Bank Group, 2022). Even exemptions from prospectus requirements for issuances to qualified investors or for issuances below a certain threshold in some jurisdictions do not make it easy enough for SMEs to participate in IPO processes (World Bank Group, 2022). Most of these companies are unable to obtain the credit they need from banks, as they are unable to provide the collateral that banks require. Furthermore, the lack of a stable cash flow necessary for credit repayments, particularly among newly established companies, means that the reliability sought by banks in order to grant credit cannot be provided, thus resulting in a lack of access to this source of financing (Korkmaz, 2020). Given these challenges, crowdfunding costs are considerably lower, and this, combined with the efficiency of using internet platforms to promote offerings, can make raising equity a realistic option for SMEs (World Bank Group, 2022).

Given that equity crowdfunding results in the issuance of equity to investors, compliance with general securities legislation is typically required for both the platforms and the SME issuers in most jurisdictions. As a result, there are strict constraints on the entities permitted to intermediate the investment (i.e., the platforms), the parties allowed to issue the securities (the fundraisers, here the SME), and the conditions under which they may do so as well as on the types of investors eligible to invest in this form of equity (World Bank Group, 2022). Markets characterized by low returns on traditional investments and high acceptance of online social networks for conducting business appear to be more advantageous for the implementation of equity crowdfunding. Only regulatory frameworks that legitimize the business, protect investors adequately, and safeguard minority shareholders can create a facilitative environment and reduce legal uncertainty (World Bank Group, 2022).

With the growing acknowledgment of equity crowdfunding's potential for SMEs, regulators across various jurisdictions are exploring methods to foster its development by reducing regulatory costs without undermining investor protection. Existing securities legislation is widely regarded as being disproportionate and inappropriate for minor activities such as crowdfunding. This phenomenon is prompting a growing number of regulators to adopt a regulatory regime for equity crowdfunding activities (for example, Brazil, Dubai, the EU, and Malaysia) or to introduce exemptions into existing capital market regulations (Australia, U.S.) (World Bank Group, 2022).

In order to strike a balance between facilitating the development of the sector and protecting investors, it is particularly important to define the permitted activities for crowdfunding platforms, introduce licensing and minimum paid-up capital requirements for platforms, establish corporate governance and conflict of interest standards, and provide investor protection standards, as well as disclosure and reporting rules for SME issuers. Achieving this balance is imperative, as it ensures the preservation of retail investors' interests while maintaining cost-effectiveness and generating appealing investment prospects. In light of these considerations, a comprehensive, tiered, risk-based approach should be considered for this purpose (World Bank Group, 2022).

In Türkiye, crowdfunding is regulated by the Communiqué on Equity-Based Crowdfunding (III-35 35/A.1) issued by the Capital Markets Board on October 3, 2019. While this Communiqué only permitted equity-based crowdfunding, the Communiqué on Crowdfunding No. III-35/A.2, adopted on October 27, 2021, repealed the first Communiqué and stipulated that crowdfunding can be based on both equity and debt (Kahveci, 2023).

According to the latter Communiqué, crowdfunding platforms must be listed by the Capital Market Board in order to operate (Art. 5/1). The platform must be established as a joint stock company and have a minimum paid-in capital of 1 million Turkish Liras (Art. 5/1). An investment committee should be established within the platform, and incoming projects should be evaluated by this committee.

According to the Communiqué, equity-based crowdfunding can only be carried out in exchange for shares to be issued through a capital increase; it cannot be carried out in exchange for the sale of existing shares (Art. 16). Debt-based crowdfunding, on the other hand, can only be carried out in exchange for debt instruments, and cannot be carried out through a loan or lending agreement or any other agreement that creates a debt relationship, or in exchange for any capital market instrument other than debt instruments (Art. 19).

In both types of crowdfunding, the funds collected from investors must be blocked in accounts opened on behalf of the platform until the end of the campaign period, and after the campaign period is completed by reaching the targeted fund amount, they must be blocked in accounts opened on behalf of the venture company and transferred to the venture company. If the targeted amount is not met during the campaign period, the funds collected will be returned to the investors. Identity verification of the people who are members of the platform is required (Kahveci, 2023). Investors can invest up to a maximum of 50.000 Turkish Liras or 10% of the investor's annual net income declared to the platform, provided that it does not exceed 200.000 Turkish Liras in a calendar year. Furthermore, the entire investment cannot be made in a single project, and the maximum investment amount that an investor can make in a project is 20,000 Turkish Liras. These thresholds are updated annually by the Capital Markets Board (SPK), and for 2026, the maximum investment amount that can be made through debt-based crowdfunding within a calendar year has been updated to 500,000 Turkish Lira, and the maximum investment amount that can be made based on income declaration has been updated to 2,000,000 Turkish Lira. The maximum investment amount an investor can make in a project has also been updated to 200,000 Turkish Liras (Sermaye Piyasası Kurulu, 2025). These amounts are relatively low. This shows that the main function of crowdfunding in Türkiye is to serve as an alternative solution that, on the one hand, prevents investors from making high-risk investments and, on the other hand, meets the funding needs of entrepreneurs by bringing together small investments.

Crowdfunding in EU law is regulated by Regulation (EU) No. 2020/1503 of the European Parliament and of the Council of October 7, 2020, on European crowdfunding service providers for businesses. The Regulation states that crowdfunding is intended to be an alternative form of financing for start-ups and SMEs based on small investments (Kahveci, 2023).

The EU Directive provides that crowdfunding may involve either facilitating the granting of loans or the issuance of transferable securities (Art 1/a). According to Directive 2014/65/EU, transferable securities are shares in companies and other securities equivalent to shares in companies, partnerships or other entities, and depositary receipts in respect of shares; bonds or other forms of securitized debt, including depositary receipts in respect of such securities; and any other securities giving the right to acquire or sell any such transferable securities or giving rise to a cash settlement determined by reference to transferable securities, currencies, interest rates or yields, commodities or other indices or measures. A loan, on the other hand, is defined as an agreement in which an investor provides a certain amount of money to a project owner for a specified period, and the project owner is unconditionally required to repay this amount, along with interest, according to an agreed repayment schedule (Art. 1/b). (Art. 1/b). Since Turkish law explicitly states that loans other than debt instruments cannot be granted, it can be said that the scope of the regulation in EU law is broader. However, the EU regulation stipulates that crowdfunding service providers must verify the absence of any criminal record related to commercial offenses on the part of the project owner. Additionally, they are obligated to gather information from prospective investors regarding their experience, investment objectives, financial status, and their comprehension of the inherent risks associated with investing in general and with specific investment types offered on the platform. Furthermore, they are obligated to provide risk disclosures. These requirements are designed to mitigate the risks associated with crowdfunding (Kahveci, 2023).

2.5 Other Instruments

As mentioned above, one of the main challenges SMEs face in accessing finance is that banks find them unattractive, because providing small loans requires almost the same amount of work as providing larger loans. Several fintech solutions have emerged to address this problem. New automation processes and financial products help reduce transaction costs and speed up both borrowing and equity fundraising by automating customer onboarding, underwriting, due diligence, and collections, while also lowering regulatory compliance costs (World Bank Group, 2022).

Another major barrier to SME financing is *information asymmetry*, which arises from the lack of reliable financial and credit data. Many SMEs do not have proper financial statements, or they may be unregistered and lack official documents that larger firms usually have. As a result, financiers find it difficult to assess and monitor their credit risk (World Bank Group, 2022). SMEs also tend to be younger businesses with little or no financing history in traditional markets, which results in limited publicly available credit records. This further weakens lenders' ability to evaluate risk. Traditionally, this information gap is compensated for by requiring higher levels of collateral (World Bank Group, 2022). However, the value of collateral and its usefulness in improving access to finance depend heavily on whether there are effective systems to identify, value, register, monitor, and enforce collateral rights (World Bank Group, 2022).

Fintech solutions have also been developed to help reduce information asymmetry in SME financing. These solutions use alternative data sources, big data analytics, artificial intelligence, and machine learning to provide lenders with additional information that supports more accurate credit risk assessments. Data from payment processors (credit card clearing companies and payment systems), e-commerce platforms, and digital banks can be used to measure cash flow, income, and repayment capacity (World Bank Group, 2022). For example, LenddoEFL, based in Singapore, was one of the first companies to use psychometric testing in credit scoring. Using more than 10,000 data points, it offers credit risk assessment services to banks and financial institutions in emerging markets for customers who may not have previously had access to loans. Following the success of alternative data-based credit scoring models, banks have started developing their own models and partnering with fintech firms offering data analytics and credit risk assessment tools (World Bank Group, 2022).

Another challenge SMEs face is that lenders often demand real estate as collateral. Therefore, improving access to finance depends on modernizing collateral frameworks and allowing the use of movable assets with clearly measurable value as collateral (Bazarbash & Beaton, 2020; Korkmaz, 2020; World Bank Group, 2022). Several fintech-driven solutions now support the use of movable assets—such as equipment, inventory, accounts receivable, payment instruments, and cash deposits—as collateral. These solutions include factoring, reverse factoring, secured revolving credit lines, merchant receivables financing, and tokenized assets (World Bank Group, 2022). While factoring and reverse factoring are traditional methods, fintech firms can enhance them by providing better collateral valuation, improved credit risk assessment, and continuous monitoring (World Bank Group, 2022). In addition, digital platforms that record transactions, links to independent databases to verify collateral (such as tax authorities, payment processors, and bank accounts), Internet of Things devices to track asset status, and smart contracts to automate agreements all contribute to making asset-based financial products more accessible and affordable for both SMEs and financiers (World Bank Group, 2022).

3 Conclusion

Financial inclusion is a hot topic that refers to consumers and businesses being able to access and use financial products and services that meet their needs. It is well known that consumers and SMEs, particularly in developing countries, face serious difficulties in accessing finance today. Fortunately, fintechs have a positive impact on changing this situation. Fintechs can help close the financing gap for SMEs and consumers by improving access to traditional financial institutions or direct financing through new digital products and process automation.

Since traditional institutions operating in the financial sector are subject to highly detailed regulations in many countries, it is an important question whether fintech companies, which want to operate in the same sector but are not yet as institutionalized as traditional institutions because they are still developing, will also be subject to the same regulations. The development of FinTechs can encourage competition and innovation in the financial services sector, enabling millions of individuals without bank accounts or with insufficient bank accounts, as well as SMEs experiencing difficulties due to a lack of access to finance, to participate in the formal financial system, thereby promoting greater financial inclusion and socioeconomic empowerment. On the other hand, since the confidentiality of customers' financial data and the protection of their assets are extremely important, it is crucial that institutions operating in the financial sector have an adequate technical infrastructure and reliability to ensure this. This is usually achieved through strict legal requirements. To be honest, whether or not to introduce a flexible legal framework for FinTechs is a matter of legal policy preference. While emerging economies with access to finance problems often prefer to adopt a flexible legal framework that encourages the growth of fintechs, states without access to finance problems may be more reluctant in this regard.

The study provides a comparative analysis of the legal regulations introduced in Turkish and EU law regarding the most common fintech institutions, products, and solutions today, while also highlighting noteworthy regulations and practices in various European countries. In this context, the legal status of e-money and payment institutions, digital banks, open banking, and crowdfunding has been examined.

To summarize the results obtained from these comparisons, first of all, legal regulations in Türkiye generally follow EU regulations, although some deviations have been identified. For example, the regulatory framework for e-money and payment institutions in both systems is quite similar. Both legal systems officially recognize these institutions and give them some rights and authorities. The most significant difference we have identified in this regard is that, under Turkish law, e-money and payment institutions are not authorized to provide any form of credit service, whereas under EU law, these institutions can provide credit services limited to their payment service activities. This indicates that fintechs are not yet permitted to engage in credit activities under Turkish law. However, this gap is filled by regulations concerning digital banks, which exist in Turkish law but not in EU law.

Turkish law has a specific regulation on digital banks, which can provide services solely through electronic banking service distribution channels without any obligation to open branches. Since there is no such regulation in EU law, banks wishing to operate without branches must also obtain a standard banking license. Turkish legislation on digital banks clearly states that digital banks must target disadvantaged consumer groups, such as SMEs, housewives, students, and young people under the age of 18 who have difficulty accessing finance, and develop products and services to meet this need. These products and services may include credit cards, cash withdrawals, credit deposit accounts, and small loans. The aim of this rule is to strike a balance between meeting the urgent or small-scale financing needs of disadvantaged groups and preserving the financial stability of digital banks. We believe this is a crucial step toward achieving financial inclusion.

Open banking is an important tool for Fintech companies to provide banking services, and in recent years, open banking applications have been supported by regulations. In this regard, the principles adopted in Turkish law and EU law are almost identical. Both legal systems require banks to share data with authorized institutions upon customer request. Currently, there are two open banking applications included in the legal regulations: account information services and payment initiation services. Thanks to open banking, Fintechs can offer these services to their own customers through banks. The only difference between the EU and Turkish legal systems is that the regulations in Turkish law are more detailed, and there are particularly detailed regulations regarding banking as a service (BaaS).

Finally, crowdfunding, which enables the collection of small amounts of equity capital from many retail investors ("crowd") by issuing shares through an online platform, thereby facilitating access to finance, particularly for SMEs, is subject to almost the same principles in EU law and Turkish law, with some minor differences. In both legal systems, crowdfunding can be equity-based or debt-based. Under Turkish law, equity-based crowdfunding can be carried out in exchange for shares to be issued through a capital increase; it cannot be carried out in exchange for the sale of existing shares. Debt-based crowdfunding, on the other hand, may only be carried out in exchange for debt instruments and cannot be carried out in exchange for a credit or loan agreement or any other agreement creating a debt relationship, or in exchange for any capital market instrument other than debt instruments. Under EU law, crowdfunding can be carried out in exchange for the granting of credit or the issuance of transferable securities. According to the EU Directive, transferable securities may include, but are not limited to, company shares, other securities equivalent to shares, depository receipts, bonds, and a wide variety of other types. On the other hand, a debt instrument might also be in the form of a loan. Since Turkish law explicitly states that only shares that can be issued through a capital increase can be offered for crowdfunding and that loans other than debt instruments cannot be offered, it can be said that the scope of the EU regulation is broader. In contrast, EU law has introduced quite detailed regulations for crowdfunding platforms regarding the verification of project owners' reliability, thereby attempting to establish a balance in this regard.

All of the regulations we have examined above show that the regulatory framework is also quite similar, given that the development of fintechs is parallel and similar types of products and services are available in the EU and Türkiye. The fundamental aim of the regulatory framework is to support fintechs, which offer a significant opportunity to ensure financial inclusion, while establishing a reliable system to protect the data and assets of fintech users. To date, the fintech sector has been developing at a rapid pace and continues to develop new products and services, so it can be said that legal regulations are one step behind. This is ultimately a legal policy choice because while some legal systems have introduced general, framework-level regulations concerning fintech, thereby aiming to encourage the development of new fintech products and services, other legal systems are attempting to introduce regulations aimed at supporting the benefits and preventing the risks of these products and services after observing their potential benefits and risks for a period of time. It can be said that both the Turkish and EU systems ultimately adopt the second of these approaches.

References

Akıncı, İ. (2024). *Kripto Varlıkların Mukayeseli Hukuk Açısından İncelenmesi ve İncelemelerin Türk Hukukuna Uygulanabilirliği.* Oniki Levha Yayıncılık.

Arıcı, M. F. (2020). Bankacılık Hukukunda Yeni Ödeme Hizmetleri ve Ödeme Hizmetleri Rejiminden İstisna Edilen Bazı Faaliyetler. In A. Kaya, B. İ. Engin, A. Paslı, Ş. İ. Aşıkoğlu, & E. Oğuz (Eds.), *Türk Hukukunun Avrupa Birliği Hukukuna Uyumu, Özel Hukuk- Acquis Communautaire'in Alınması -Açıklamalar, Değerlendirmeler, Öneriler* (pp. 297–337). İstanbul Üniversitesi Yayınevi.

Arner, D. W., Buckley, R. P., & Zetzsche, D. A. (2022). Open banking, open data, and open finance: Lessons from the European Union. In O. Banking (Ed.), *Jeng, L* (pp. 147–172). Oxford University Press.

Bazarbash, M., & Beaton, K. (2020). *Filling the gap: Digital credit and financial inclusion* (IMF Working Paper, WP/20/150). Monetary and Capital Markets Department.

BEUC (the European Consumer Organisation). (2023). *Payment Services: BEUC's recommendations on the Payment Services Regulation and the Payment Services Directive 3.* https://www.beuc.eu/sites/default/files/publications/BEUC-X-2023-123_Payments_Services_Recommendations%20on%20the%20PSR-PSD3.pdf

Bispo, A. A. (2020). Open banking and the decentralization of payments in the EU: Untwining the PSD2. *European Data Protection Law Review (EDPL), 6*(4), 586–592.

Cardinal, D., & Thomas, N. (2022). Data access technology standards: A history of open banking data access. In O. Banking (Ed.), *Jeng, L* (pp. 91–106). Oxford University Press.

Casolaro, A. M. B., Rauber, G. N., & Monteiro de Lima, U. (2025). Open banking: A systematic literature review. *Journal of Banking Regulation, 26*, 340–355.

Christensen, L. D. (2025). *EU payment services: Regulation and innovation.* Oxford University Press.

Çeşit, C. K. (2021). Ödeme Hizmetlerindeki Rekabette Yeni Dönem: Madalyonun İki Yüzü. In K. C. Sanlı & D. Alma (Eds.), *Uygulamalı Rekabet Hukuku Seminerleri 2020* (pp. 252–260). Oniki Levha Yayıncılık.

Divissenko, N., & D'Amico, A. S. (2025). Implementing interoperability: Regulating innovation and competition under PSD2 and DMA. *European Journal of Risk Regulation*, 1–16. https://doi.org/10.1017/err.2025.10061

Erel, I., & Liebersohn, J. (2022). Can FinTech reduce disparities in access to finance? Evidence from the Paycheck Protection Program. *Journal of Financial Economics, 1*(146), 90–118.

European Commission. (June 28, 2023a). *Modernising payment services and opening financial services data: New opportunities for consumers and businesses* [Press Release] https://ec.europa.eu/commission/presscorner/detail/en/ip_23_3543

European Commission. (June 28, 2023b). *Payment services: revised rules to improve consumer protection and competition in electronic payments* [Questions and Answers] https://ec.europa.eu/commission/presscorner/detail/en/qanda_23_3544

Gürses, D. (2025). *Hukuki Açıdan Açık Bankacılık ve Dijital Bankalar*. Oniki Levha Yayıncılık.

Herrera, L. A. (2025). PSD3 and the regulation on payment services in the context of crypto assets as a means of payment. In C. P. Sempere (Ed.), *Governance and control of data and digital economy in the European single market law, legal framework for new digital assets, identities and data spaces* (pp. 373–394). Springer, Open Access. https://doi.org/10.1007/978-3-031-74889-9_16

Huang, Y., Zhang, L., Li, Z., Qiu, H., & Sun, T. (2020). *Fintech credit risk assessment for SMEs: Evidence from China* (IMF Working Paper No. 20/193). Asia Pacific Department.

Iyelolu, T. V., Agu, E. E., Idemudia, C., & Ijomah, T. I. (2024). Legal innovations in FinTech: Advancing financial services through regulatory reform. *Finance & Accounting Research Journal, 6*(I. 8), 1310–1319.

Jeng, L. (2022). Inception to open banking. In O. Banking (Ed.), *Jeng, L* (pp. 1–11). Oxford University Press.

Kahveci, D. (2023). *Sermaye Piyasası Hukukunda Kitle Fonlama Platformları.* Oniki Levha Yayıncılık.
Karakoç, E. (2024). *Açık Bankacılık Uygulamalarının Hukuki İncelemesi.* Oniki Levha Yayıncılık.
Korkmaz, Ö. (2020). *Amerikan Hukuku ile Karşılaştırmalı Paya Dayalı Kitle Fonlaması.* Oniki Levha Yayıncılık.
KPMG Türkiye. (2024). *Bankacılık Sektörel Bakış.* https://assets.kpmg.com/content/dam/kpmg/tr/pdf/2024/10/Bankac%C4%B11%C4%B1k%20Sekt%C3%B6rel%20Bak%C4%B1%C5%9F%202024%20v5.pdf#page=3.99
Lupo-Pasini, F. (2021). Financial inclusion and the "war for cash." *Law and Contemporary Problems, 84*(1), 17–34.
Miglionico, A. (2022). Digital payments system and market disruption. *Law and Financial Markets Review, 16*(3), 181–196.
Miriyeva, N. (2022). European payments in the digital age. *ELTE Law Journal, 2022*(2), 35–60.
Perçin, Ö., Deregözü, R., İlik, H., Varlı, A.T., Eren, E., Erçetin, P., & Gürses, D. (2023a). Elektronik Para Kuruluşları ve Ödeme Kuruluşları. In Ö. Perçin (Ed,), *Sorularla Fintek* (pp. 1–76). Oniki Levha.
Perçin, Ö., Deregözü, R., & Gürses, D. (2023b). Açık Bankacılık, Servis Modeli Bankacılık (BAAS) ve Dijital Bankacılık. In Ö. Perçin (Ed.), *Sorularla Fintek* (pp. 77–122). Oniki Levha.
Pouelle, J. B., Kannan, A., Spitz, N., Kahn, S., & Sotiropoulou, A. (2024). *EU banking and financial regulation.* Edward Elgar Publishing.
Sermaye Piyasası Kurulu. (2025, December 31). *Duyuru ve İlke Kararları, 2025/68* [Press Release]. https://spk.gov.tr/data/695546ca8f95db281807a9c3/2025-68.pdf
Sünbül, N. (2023). *Finansal Teknoloji Ortaklıkları ve Ödeme Hizmetleri Rejimi.* Oniki Levha Yayıncılık.
TCMB (Türkiye Cumhuriyet Merkez Bankası). (December 26, 2022). *Ödeme ve Elektronik Para Kuruluşlarının FAST Sistemine Katılımına İlişkin Basın Duyurusu* [Press Release]. https://www.tcmb.gov.tr/wps/wcm/connect/tr/tcmb+tr/main+menu/duyurular/basin/2022/duy2022-53#:~:text=2023%20y%C4%B11%C4%B1nda%20%C3%96deme%20ve%20Elektronik,y%C4%B11%C4%B1n%C4%B1n%20ilk%20%C3%A7eyre%C4%9Finde%20kullan%C4%B11maya%20ba%C5%9Flanacakt%C4%B1r
Wamba-Taguimdje, S. L., & Kamdjoug, J. R. K. (2024). Mobile payments and money technologies in sustainable development: A systematic literature review and computer-assisted interpretive analysis. *Information Technology for Development, 31*(2), 435–472.
Wolberg-Stok, A. (2022). Open banking ecosystem and infrastructure: Banking on openness. In L. Jeng (Ed.), *Open banking* (pp. 13–30). Oxford University Press.
World Bank Group. (2022). *Fintech and SME finance: Expanding responsible access.* International Bank for Reconstruction and Development/The World Bank.
Zunzunegui, F. (2018). *Digitalisation of payment services.* Revista de Derecho del Mercado Financiero (Working Paper No. 1/2018). SSRN: https://ssrn.com/abstract=3264759 or https://doi.org/10.2139/ssrn.3264759

Business Responsibilities in Achieving Sustainable Development Goals Through Fintech

Özcan Serhat Cengiz

1 Introduction

"*Business responsibility*" is the new buzzword. It owes its recent popularity in part to efforts to scrap "*ESG*" and replace it with new terms.[1] Nevertheless, its actual meaning is important and conducive to yielding a significant impact. With this motivation, this contribution investigates the concept of business responsibility and its implications for sustainable development through fintech.

It first starts with visiting relevant corporate law literature. As the primary legal domain shaping the actions of corporations, corporate law mainly regulates the relationship between shareholders and directors of companies owned by them. It imposes fiduciary duties on directors with a view to credibly committing them to the interests of shareholders. This complicates responsible business efforts and, in principle, requires directors to navigate trade-offs originating from their commitment to shareholder wealth. Taking this framework as a point of departure, this subsection conceptualizes business responsibility and provides guidance on how to accommodate legal normative judgments on business responsibility.

The following subsection explains the relevant asset-pricing literature related to business responsibility. In that vein, it is shown that responsible business practices might serve the interests of both shareholders and stakeholders concurrently. Yet, it is indicated that this is a highly complicated strategy that only competent directors can benefit from.

[1] Corporations, particularly in the US, are increasingly using the term "responsible business" instead of "ESG" to avoid spotlights (Cutter & Glazer, 2024).

Ö. S. Cengiz (✉)
Leibniz Institute for Financial Research SAFE and Goethe University, Frankfurt, Germany
e-mail: cengiz@safe-frankfurt.de

B. Ş. Şahin and B. C. Cantürk (eds.), *Fintech and Sustainable Development*, Sustainable Finance, https://doi.org/10.1007/978-3-032-31020-0_6

Then, the Sustainable Development Goals ("**SDGs**") are introduced as soft-law governance instruments. Their success in reconciling diverse objectives is highlighted, and several empirical studies are reviewed to assess prevalence in practice. This subsection also identifies specific areas where the implementation and effectiveness of the SDGs could be further enhanced.

The discussion then turns to sustainability-oriented innovation, which refers to new products or processes that support social and environmental goals while engendering innovation. After explaining this concept, fintech is suggested as a convenient field of play where it is highly relevant. Because fintech companies rely heavily on digital innovation, they can develop solutions that enable "doing well by doing good", ultimately benefiting society, the environment, and their shareholders.[2] It concludes by emphasizing fintech's potential to further contribute to sustainable development and to continue flourishing as a promising sector.

2 The Blueprint of Business Responsibility

2.1 The Meaning of Business Responsibility

The notion of responsibility has been a topic of rigorous discussion in relation to businesses.[3] At first glance, responsibility resonates with an ethical context and does not connote businesses. However, as corporations have assumed increasingly important roles in society (Gatti, 2025), this paradigm shift has prompted significant reconsiderations. Consequently, interest in revisiting the foundational principles underlying corporate existence has substantially increased.[4]

The increasing interest in business responsibility has not only yielded considerable depth of analysis but also occasionally caused confusion. This is primarily caused by the lack of a mutually agreed-upon definition of "business responsibility". According to the European Commission, business responsibility is "*a concept whereby companies integrate social and environmental concerns in their business operations and in their interaction with their stakeholders on a voluntary basis*" (European Commission, 2011). According to the OECD, responsible business conduct is about "*making sure that the activities of businesses and investors are aligned with the needs of society, today and in the future*" (OECD, 2025) whereas the United Nations Industrial Development Organization explains corporate social responsibility as "*being the way through which a company achieves a balance of economic, environmental and social imperatives*" (UNIDO, 2025). These definitions feature various aspects of business responsibility. The European Commission uses binding laws as a benchmark

[2] The concept of "doing well by doing good" is explained under Sect. 2.2. For an early conceptualization in the literature, see Falck and Heblich (2007).

[3] In this chapter, the term "business" is used interchangeably with "corporation", meaning a separate legal for-profit entity with ownership and management separation.

[4] See, for example, *The Economist* (2019).

and emphasizes the voluntary nature of business responsibility, whereas the OECD includes investors in its definition, and the United Nations Industrial Development Organization prioritizes achieving a balance among economic, environmental, and social considerations in its definition. There is no mutually agreed-upon definition among academic scholars either.[5]

The lack of a mutually agreed-upon definition may, and at times does, lead to confusion, particularly in legal discussions. This is primarily because business responsibility is discussed within the context of "normative judgments", and it is not easy to discern whether such a judgment is suggested as *de lege lata* or *de lege feranda*. In other words, it is often unclear whether such a judgment claims that positive law requires companies to act more responsibly, or whether positive law itself should be formulated in a way to require companies to act more responsibly. To avoid possible confusion about the meaning of business responsibility, the following paragraphs briefly review the relevant corporate law literature, followed by a conceptualization for the purposes of this contribution.

Historically, it has been argued that corporations have a single responsibility: maximizing profits (Friedman, 1970). This fundamental assumption has led to the shareholder primacy phenomenon and has shaped the rules of corporate law in many jurisdictions. Although some countries, such as Germany and France, require directors to consider the interests of stakeholders alongside those of shareholders, even their corporate law frameworks still prioritize shareholder interests.[6]

This perspective is strongly criticized by numerous scholars. It is argued that confining discussions of corporations to a principal-agent tension oversimplifies their dynamics (Blair & Stout, 1999), and that an excessive focus on shareholder value puts pressure on directors to prioritize short-term value maximization at the expense of long-term value creation (Lund, 2021; Stout, 2012). Further, corporations' profit-maximization purpose has led them to increasingly harm others. In that vein, although corporations emerged in society to solve certain problems, they have become, on the contrary, organizations that generate new problems or contribute to existing ones (Mayer, 2024).

To mitigate these adverse impacts of shareholder primacy, it is posited that corporate law rules should be fundamentally revised. This proposition is supported by several sub-arguments. In that vein, it is suggested that demand for responsible business practices from shareholders is growing, and this growing demand should be embraced by corporate directors. This argument is based on the predicament that corporate directors should prioritize shareholder welfare over shareholder value, which can represent not only the financial but also ethical and social motivations of shareholders (Hart & Zingales, 2017, 2024).[7] In addition, institutional shareholders

[5] For an overview of different definitions of corporate social responsibility, see Kitzmueller and Shimshack (2012).

[6] For Germany, see Binder (2024). For France, see Hopt (2024).

[7] This explanation is closely related to the concept of "delegated philanthropy," which suggests that investors may be willing to forgo a portion of their profits in order to induce prosocial activities by the companies they own (Bénabou & Tirole, 2010).

are becoming more salient in financial markets as "portfolio value maximizers". As opposed to "firm value maximizers", those institutional shareholders have more incentives to ask their portfolio companies to mitigate the harm they inflict on others (Gordon, 2022). Therefore, corporate law rules should be adapted to address this structural novelty (Enriques & Romano, 2022). Ultimately, the argument follows that corporate laws should be revised to hold directors accountable not only to their shareholders but also to other stakeholders. In the European Union, these considerations were raised in a prominent report, which subsequently sparked rigorous discussions (European Commission & EY, 2020).[8]

Although the arguments against shareholder primacy have solid grounds, they are criticized for failing to offer concrete and viable solutions. First, mandating directors to serve not only shareholders but also other stakeholders are susceptible to the "too many masters" problem (Kuntz, 2024).[9] Moreover, corporate directors are business professionals, but not policymakers. Therefore, they are deemed not to have the necessary skills to identify social and environmental policies effectively, nor the democratic legitimacy to serve stakeholders other than shareholders (Edmans, 2023a; Hopt, 2024). Although it is not possible to review all the relevant literature in corporate law, it wouldn't be wrong to conclude here that this discussion is far from being settled.[10]

Against this brief background, the jurisdictions this book focuses on predominantly require directors to prioritize shareholders' interests under corporate law rules, and, at least as of now, there is no sign of a change in that regard.[11] This yields three important conclusions: First, directors will strive to maximize their companies' value to enhance shareholders' wealth. In that vein, in principle, shareholders will induce managers either to minimize the share of other stakeholders in the generated wealth or, at best, to remain indifferent in terms of contributing to their welfare. Second, companies will fail to recognize the harm they inflict on others while prioritizing the maximization of their shareholders' wealth, resulting in negative externalities. Finally, in the absence of effective limitations imposed by corporate law, regulatory

[8] For a rebuttal of this report arguing that its analyses and conclusions are fundamentally flawed, see Roe et al. (2021). For another critical approach to the said report that highlights the role of corporate governance codes, see Ferrarini (2025).

[9] In that vein, it is argued that being accountable to stakeholders in addition to shareholders would make directors accountable to "everyone" and being accountable to everyone is equal to being accountable to no one (Bainbridge, 2023) because in such a case, directors would lack the guidance about how to navigate trade-offs between the interests of various constituencies (Jensen, 2010).

[10] A recent paper brings a fresh and interdisciplinary perspective to this discussion. Accordingly, Hwang and Lund (2025) investigate non-profit enterprises in the US that, in certain cases, outperformed their for-profit peers. They posit that this can be explained by insights from behavioral economics and organizational science literature and argue that a non-profit purpose for a company can replace shareholder monitoring and align incentives across enterprises. In the same vein, another paper draws on the organizational psychology literature and emphasizes how organizational purpose can motivate employees within a firm (Ferrarini, 2025).

[11] The lack of indicators of imminent change does not corroborate the claim that the incumbent corporate law structure is the optimal one. Rather, it recognizes that it is difficult to change it due to inherent dependencies on the existing system (Lund & Pollman, 2021).

interventions will have to be designed to safeguard non-shareholder stakeholders, either by preventing corporations from producing externalities or by obliging them to internalize such externalities caused by their activities.[12]

The three conclusions outlined above regarding corporate law rules serve as a framework for the concept of "business responsibility". Accordingly, (i) businesses that contribute to the welfare of their non-shareholder stakeholders more than required by regulations and (ii) businesses that try to mitigate the negative externalities they cause (again, more than required by law) should be deemed responsible.

Obviously, such conceptualization is neither innovative nor provides a clear distinction between responsible companies and those that are not. On the contrary, it is a rather flexible benchmark. Nevertheless, it remains useful for several reasons. First of all, it clarifies that business responsibility is outside the legal domain.[13] In accordance with corporate law's mandate, directors aim to maximize shareholders' wealth, while regulations curb this ambition by imposing certain limitations. It is the directors' legal duty to abide by these imperatives (Kuntz, 2024). However, business responsibility means beyond compliance with the law. Therefore, it corresponds to voluntary corporate actions that serve the interests of non-shareholder stakeholders, even though they are not undertaken as a response to legally binding obligations.

Secondly, responsibility is inherently a relative concept. As such, the meaning of being responsible is highly dependent on the social values prevalent in society at a given time and is subject to constant change. For instance, views relating to corporations in the US as agents in society shifted significantly in the twentieth century in response to two major streams of events. First, the 1929 crisis led to the proliferation of the view that corporations have duties not only to their owners but also to other stakeholders (Lund, 2025). Another turning point was in the 1970s and 1980s. Influential papers by economists such as Milton Friedman, Michael Jensen, and William Meckling paved the way for a shareholder-centric perspective to replace the then prevailing managerial-freedom-based views (Lipton, 2025; Lund, 2025). The diffusion of shareholder-centric views of corporations was further reinforced by the wave of hostile takeovers in the 1980s (Lund, 2025). These insights from US corporate law history show that societal paradigm shifts can lead to significant changes in companies' images. Therefore, business responsibility must be a flexible concept that captures and adapts to such shifts in the relevant jurisdiction, time, and context.

More importantly, this conceptualization enables further analytical legal mapping of normative judgments related to business responsibility. Legal scholars often raise

---

[12] In this context, regulation refers here to binding legal norms originating from areas of law other than corporate law, such as environmental or employment law.

[13] By "legal domain" it is referred here to the set of rules that are binding in nature and imposes sanctions in the event of its violation. Under soft law rules, corporations are expected to behave "responsibly" by observing non-shareholder stakeholders. However, soft law rules do not impose a legal obligation on the companies to deliver concrete outcomes and do not sanction irresponsible actions. Therefore, they are outside the scope of legal domain for the purposes of this conceptualization. For a further explanation of soft laws within this context, see Sect. 2.3.

concerns about corporations' sustainability performance and argue that corporations should act more responsibly.[14] As discussed above, from a legal perspective, such a normative judgment can have two meanings. The first possible meaning can be descriptive, in the sense that positive law requires corporations to act more responsibly, and corporations should comply with such laws. As discussed above, business responsibility is a concept that falls outside the legal domain, as it describes corporate actions undertaken voluntarily, without any legal obligation. Therefore, this meaning is eliminated.

The second possible meaning is a legal normative one. It can be argued, *de lege feranda*, that positive law should be amended to establish a legal obligation to act or refrain from acting in a certain way, thereby requiring corporations to become more responsible. This can be a fairly valid legal argument, which should be subjected to an in-depth regulatory feasibility assessment. Accordingly, the first step in such an assessment is to identify the relevant non-shareholder stakeholders that corporations are required to observe. The second step is to assign the relevant field of law that seeks to protect the identified stakeholder's interests as a policy goal. Given that normative judgments alike are almost always relevant for more than one field of law, such an assessment should be conducted jointly for all relevant fields of law.

Once the legal mapping is complete and the relevant fields of law are identified, the next step is to conduct a regulatory assessment to finalize the normative judgment. While doing so, it should be borne in mind that the goal is to deploy the most effective and efficient policy tool(s) that would yield the pursued outcome. In that vein, one option is opting for the traditional command-and-control regulation, which can be summarized as state-imposed requirements mandating uniform precautionary or technological measures (Coglianese & Kagan, 2007). A second option is to adopt a meta-regulation approach. Meta-regulation refers to a system in which external regulators intentionally encourage organizations to design and implement their own internal control mechanisms (Coglianese & Mendelson, 2010). Another option is self-regulation. In this alternative, "targets" of regulations impose regulations on themselves without any formal legal intervention by regulators (Coglianese & Mendelson, 2010). Of course, these three forms of regulation are neither mutually exclusive nor an exhaustive list for policy considerations. However, they remind us that it is not a binary choice between regulating and not regulating, and there are several means in the policymaker's toolkit.

When considering various policy options, it is essential to recognize the strengths and weaknesses of policy tools in the context of business responsibility. Unlike the first one, in meta and self-regulation, irresponsible behavior does not constitute a breach of legal duties or cause legal liability. This can be deemed an insufficient intervention in corporations' incentive structures. However, meta-regulation and self-regulation are not toothless. The increasing demand for responsible behavior from

[14] For an example pointing at banks and their "responsibilities", see Macchi (2025). "*Banks' understanding and implementation of human rights responsibilities is still insufficient. A just transition is not just about greening portfolios—it's about embedding human rights at the core of climate finance.*"

investors, contractors, and consumers means that irresponsible actions can be penalized by market actors, even beyond what a typical legal obligation would entail. Ultimately, legal normative judgments regarding business responsibility should be analyzed with consideration for all these dynamics.

This analytical legal mapping of normative clauses yields a crucial conclusion. From a regulatory perspective, corporations cannot be expected to deliver on moral values. They are distinct from individuals in society, devoid of moral independence (Osterloh-Konrad, 2024), and owe their existence to the motivation to generate profits for their shareholders. Therefore, instead of "accusing" corporations of not being responsible enough, the focus should be placed on their incentive structures and on identifying the most effective and efficient policy instruments capable of influencing them.

To conclude this subsection, business responsibility refers to voluntary corporate actions that either contribute to the welfare of non-shareholder stakeholders or mitigate the externalities it causes. This conceptualization helps define the boundaries of responsible corporate action and provides a legal analysis of normative judgments regarding business responsibilities. For the latter, a thorough regulatory assessment should be conducted to prevent regulatory failures before introducing ambitious novel interventions.

2.2 *The Making of Business Responsibility*

Given that business responsibility is only voluntary and corporate law rules hold directors accountable vis-à-vis shareholders in that they will maximize their wealth, why would directors engage in responsible corporate actions after all? One strand of literature considers responsible business as agency costs and a way to "shirk". In other words, by engaging in responsible business activities, directors violate their fiduciary duties and undermine corporate value (Lund, 2021).[15] This approach is arguably falling short of providing a comprehensive explanation. The focus here will be on the most relevant alternative explanation, which perceives being responsible as a business strategy. This description, also known as "enlightened shareholder value

[15] Indeed, there is empirical evidence suggesting that corporate directors opportunistically use corporate donations for their own private benefits at the cost of shareholder value. See Masulis and Reza (2015). For another influential paper in a similar direction, see Bebchuk and Tallarita (2020). The concern that fiduciaries would misuse investors' money was the underlying suggested justification for prohibiting asset managers in certain states in the US from "ESG investing", who are subject to similar fiduciary duties vis-à-vis their investors. A recent report indicates that 11 anti-ESG bills were passed in 10 states across the US in 2025 as of the date of the report (Gibson et al., 2025). However, it is also plausibly argued that ESG investing is not in conflict with fiduciary duties under US laws when employed with a risk-return perspective. For an investigation of fiduciary duties of asset managers and ESG investing under US laws, see Schanzenbach and Sitkoff (2020). Further, for an investigation of fiduciary duties of corporate directors and climate change under US laws, see Condon (2025) and Barker et al. (2021).

theory",[16] reconciles business responsibility with shareholder wealth maximization by asserting that socially responsible behavior can ultimately benefit shareholders by creating value (Bénabou & Tirole, 2010). How this can be realized in practice is succinctly explained by Cornell and Damodaran (2020) (Fig. 1).

This chart summarizes the value drivers of corporations according to the capital asset pricing model. As such, any exogenous factor that affects any of the elements of this chart would impact a company's value and, in turn, the return to its shareholders. This yields an important conclusion: Unlike the traditional approach to corporate business responsibility, which views responsibility as costly, there need not be a trade-off between responsible business actions and shareholder wealth. If a "responsible" corporate action has an impact on any element above and ultimately increases corporate value, such action can serve its shareholders while also benefiting other stakeholders.

As an example, it is empirically well-established that corporations with more satisfied employees generate higher earnings,[17] which can be derived from either higher revenue or better operating margins. This suggests that sharing generated wealth with employees, as a non-shareholder stakeholder, could ultimately benefit shareholders in the long term. As another example, the cost of energy from specific renewable energy sources has been steadily decreasing.[18] Therefore, switching from fossil fuel-based energy sources to renewable energy sources would not only shrink carbon footprints but also likely reduce energy costs. This would correspond to an increase in operating margin in the value chart, ultimately to an increase in the firm's value.

Two important actors within this context are investors and consumers. First, investors are increasingly incorporating responsibility criteria into their financing conditions. This applies to both capital markets and bank loans. This paradigm shift can derive from multiple incentives. Firstly, as sustainability-related risks (e.g., climate and regulatory risks) become more material, lenders and investors are increasingly willing to engage with more responsible companies to mitigate their exposure to the relevant risks. In addition, by financing responsible businesses, they enhance their own reputation and signal to the market that they are acting responsibly, thereby attracting other responsible investors. Finally, altruistic investors might prefer to finance responsible businesses for psychic motivations. Regardless of the underlying investor motivations, companies can finance their activities at a lower cost of funds by acting more responsibly.

Secondly, consumers are now more concerned about responsible businesses. Customers, particularly young ones, prefer to purchase goods and services offered

[16] For a brief explanation about the emergence of and counterarguments against enlightened shareholder value theory, see Ferrarini (2025) and Lund (2020).

[17] This is documented by numerous studies. However, for a seminal paper, see Edmans (2011).

[18] For example, in 2021, the cost of electricity derived from solar panels and wind turbines per kilowatt-hour in Europe was reported to be four to six times lower compared to the cost of fossil fuels in 2022 (Osman et al., 2023).

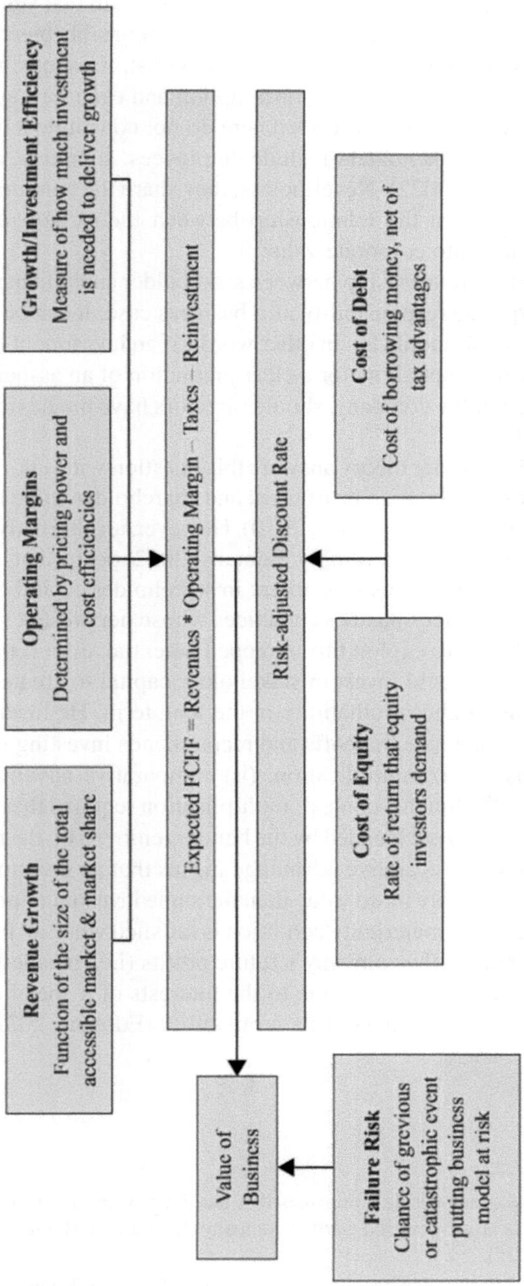

Fig. 1 Firm value drivers under capital asset pricing theory (Cornell & Damodaran, 2020)

by more responsible companies (Hart & Zingales, 2024). This incentivizes companies to become more responsible to attract more consumers. In that vein, responsible businesses differentiate their products from those of irresponsible businesses and are rewarded by higher consumption (Hart, 1995). In contrast, irresponsible businesses are punished by consumers through less consumption and even through boycotts.[19]

Naturally, the two stakeholders described here do not constitute a full list. Other relevant non-shareholder stakeholders include employees, suppliers, communities, and governments (Edmans, 2022). Nevertheless, they share the common feature that corporate managers can turn the relationship between their companies and non-shareholder stakeholders into corporate value.

Having established the relationship between stakeholder capitalism and corporate value and framed corporate responsibility as a business case, it can be further questioned how certain this link should be. In other words, if an investment in stakeholder capital by a company does not allow for a clear prediction of an associated increase in the net present value of the company, should directors have the flexibility to invest in stakeholder capital?

The enlightened shareholder theory answers this question with a no, since it views stakeholder capitalism merely as an instrument and shareholder value maximization as an end (Falck & Heblich, 2007; Jensen, 2010). However, an alternative third way is suggested by Edmans (2022). Accordingly, Edmans (2022) posits that, since enlightened shareholder theory conditions investment in stakeholder capital on net present value, it automatically excludes business practices whose net present value implications cannot be calculated. To exploit this untapped potential, even if there is no link to profitability, directors should invest in stakeholder capital to create social value, which can translate into higher profitability in the long term. He further introduces a framework on how to navigate tradeoffs and recommends investing in stakeholder capital only if it satisfies (i) multiplication, (ii) comparative advantage, and (iii) materiality conditions. Within this context, multiplication requires that the cost of an investment for a corporation is exceeded by the benefits enjoyed by the beneficiary of that investment, whereas comparative advantage means that an investment in stakeholder capital creates relatively more value than the same investment made by others (Edmans, 2022). Finally, the materiality condition is satisfied when an investment has the potential to contribute to the company's future profits (i.e., business materiality) or when the firm attributes intrinsic value to the interests of a specific stakeholder (i.e., intrinsic materiality), regardless of its profitability (Edmans, 2022).[20]

---

[19] For instance, Spotify was among the companies that faced a boycott due to its founder, CEO Daniel Ek's investments in an AI-powered warfare company (Business & Human Rights Resource Centre, 2025; Snapes, 2025).

[20] See Edmans (2022) for further explanations with examples. See also (Hahn et al., 2018) for a more radical approach towards business responsibility as a business case, arguing that the tension between profitability and business responsibility should be embraced and the latter.

As a final note, the above explanations should not lead to the conclusion that being responsible systematically leads to higher stock market performance.[21] Indeed, corporate finance theory explains that markets learn about price-sensitive information and ultimately incorporate it into prices. During this learning period, investing based on such information can yield excessive returns for investors, whereas once the learning period is over, there is no room for such returns, as prices would then reflect the information (Cornell & Damodaran, 2020). Therefore, "doing well by doing good" is a complex strategy for business managers, not only because they must comply with their fiduciary duties under corporate law, but also because it is difficult to measure the materiality of being responsible.[22] Ultimately, the duty of navigating challenging trade-offs in pursuit of doing well by doing good rests on corporate directors' shoulders.

2.3 The Role of SDGs as a Soft Law Instrument in Defining Business Responsibility

It is established in the previous subsection that business responsibility is outside the legal domain and can reasonably be pursued as a business case. This framework provides a general explanation but still lacks an important element: a set of standards for becoming responsible. When there is command-and-control regulation, laws set standards for qualifications. However, in the absence of legal mandates, as is the case with business responsibility, where can one get the conditions for a qualification?

The answer to this question lies in "soft law."[23] Soft law has been an influential tool for pursuing social and environmental goals. It is preferred by states over "hard law" such as binding treaties for several reasons. First of all, it is flexible. Unlike hard law, soft law provides a general framework so that it can be adopted by actors in different jurisdictions under different circumstances, both at the development and enforcement phases (Nolan, 2013). Moreover, it is relatively easier and less costly to agree on soft law, particularly in international governance, as it is ultimately non-binding (Nolan, 2013).

---

[21] For a recent literature review on responsibility and financial performance, see Liang et al. (2025). An influential empirical paper argued that companies that invested in certain financially material ESG aspects, as outlined by the Sustainability Accounting Standards Board, outperformed their peers (Khan et al., 2016). However, it was later shown by another paper that this indication vanishes under different modelling preferences (Berchicci & King, 2022; Edmans, 2023a). Therefore, current empirical literature fails to find convincing evidence, indicating a systematic link between ESG and financial performance.

[22] As expressed by Garst et al. (2022), identifying the material ESG aspects for a company is an art rather than a science.

[23] Although its definition is highly controversial (Shaffer & Pollack, 2010), soft law refers here to the rules introduced by governmental or non-governmental national or international organizations that are non-binding in nature.

Although it lacks legal enforceability, soft law serves as a highly functional instrument for influencing corporate behavior for several reasons. First, corporations, directors, and investors who are genuinely committed to acting responsibly may lack the expertise or information necessary to identify appropriate social and environmental objectives, particularly those related to externalities (Edmans, 2023a, 2023b). Soft law frameworks, predominantly developed by national policymakers and international organizations, offer valuable guidance by clarifying which goals should be pursued and how they should be implemented.

Another important function of soft law lies in its capacity to enhance comparability among corporations. In the absence of binding regulatory standards, it becomes difficult to distinguish actual responsible corporations from those that merely appear so. By establishing a common set of benchmarks, soft law facilitates meaningful comparison across firms. As noted in the previous subsection, attracting consumers and investors constitutes a key channel through which responsible business practices can be transformed into added corporate value. Such value creation, however, is possible when corporations credibly demonstrate their responsible behavior through disclosure to the outside world. Such a demonstration should take place through disclosure in a standardized and comparable manner so that investors can discern responsible businesses (Christensen et al., 2021). In that vein, soft law serves both as a benchmark for responsible businesses and as a means to standardize disclosures. For example, one of the most common portfolio-construction methodologies employed by institutional investors is "norm-based screening," in which investment decisions are based on whether an investee company has adopted a soft law framework (Ferrarini, 2025). By disclosing adherence to generally accepted, credible soft law frameworks, companies can attract significant capital from such investors.

These explanations provide a helpful framework for understanding SDGs as a form of soft law, highlighting their fundamental features. Within this context, the SDGs are a non-binding set of goals that provide a comprehensive framework for both industrialized and developing countries while also recognizing generous leeway for states in their implementation (Biermann et al., 2017). They were introduced by the United Nations ("**UN**") after the time frame for Millenium Development Goals was reached in 2015. It has universality as one of the underlying core themes (Long, 2015). These goals were developed through a highly inclusive process (Arora-Jonsson, 2023) that is deemed to successfully integrate economic, social, and environmental goals (Koehler, 2016). The ultimate version compiled 17 goals and 169 targets, accompanied by indicators to measure their level of accomplishment.

SDGs are distinct standards because they not only set goals, targets, and indicators, but also provide guidance on how to pursue them. As discussed before, fulfilling business responsibility is viable to the extent that it can be pursued as a business case, which is challenging. To succeed in this, the UN employs various initiatives. One of them is SDG Impact Standards. This initiative aims to help enterprises, as well as bond issuers and private equity funds, identify relevant SDGs for their businesses and guide their decision-making (SDG Impact, 2023). Another initiative is the UN Global Compact, which, among others, helps enterprises integrate SDGs into their businesses (UN Global Compact, 2025a). Finally, the SDG Action Manager Tool offered by the

UN enables corporations to set goals based on their specific circumstances, track their progress against benchmarks, and show their progress (UN Global Compact, 2025b). While doing this, the UN also offers various training programs for professionals, such as the UN Global Compact Academy, SDG Ambition Accelerator, and the Young SDG Innovators Programme, to guide corporate leaders in managing their operations sustainably (UN Global Compact, 2025b).

Overall, the UN has been working to promote the use of SDGs among private actors. Literature on international governance explains that private actors are an important part of such initiatives (Cashore, 2002; Gadinis & Fregoni, 2025) and they are conducive to delivering results to the extent that they are welcomed by industry (Gadinis & Fregoni, 2025).[24] Therefore, such efforts by the UN are of utmost importance. However, whether these efforts induced private actors to pursue the SDGs should be judged by empirical research. In that vein, one way to understand whether and how corporations accommodate SDGs is to examine their corporate reports and disclosures. Hummel and Szekely (2022) investigate this for the first three years of SDGs from 2015 to 2018. They analyze annual reports and find a significant increase in the number of companies in their sample reporting on SDGs, from 15 to 58%. Thammaraksa et al. (2024) et al. confirm that this increasing trend continued between 2019 and 2021 for their sample.

Other studies investigate the heterogeneity of SDG reporting among corporations across several factors, including sector and geographic location. Arena et al. (2023) find that among 75 companies in their sample, all oil and gas companies, 47 (62.6%), referenced the SDG framework in their 2018 reports. Erin and Bamigboye (2022) find that, except in South Africa, SDG reporting levels are considerably low across African countries.[25] Finally, Elalfy et al. (2021) finds that SDG reporting practices are more common for corporations that are (i) bigger, (ii) listed, and (iii) operating in industries with higher sustainability impacts. Another paper extends this research's findings and concludes that compliance with the Global Reporting Initiative's guidance for SDG reporting varies across sectors, with the automotive sector the most compliant and corporations in the technology sector the least compliant (Lukács et al., 2025).

The second important aspect concerns the real-world impact of SDGs, as measured by companies' sustainability performance. This is a relatively challenging aspect to explore empirically due to strong endogeneity concerns. However, it is still possible to determine whether there is at least a correlation between SDG reporting and the delivery of SDG goals.

Empirical papers exploring this relationship yield mixed results. On the one hand, several scholars find an association between SDG reporting and improved sustainability performance. For example, Beretta et al. (2025) posit a relationship between environment-focused SDG reports and environmental performance, as well

[24] As a concrete example Green (2010) investigates how Green House Gas Protocol initiative became successful in an international arena and points at the role of industry support in this success.

[25] For a strand of literature investigating the relationship between SDG reporting and country-level institutional factors, see Rosati and Faria (2019) and Galeazzo et al. (2024).

as between socially focused SDG reports and social performance for Italian companies. In a similar direction, Sharma et al. (2025) find that SDG adoption significantly enhances ESG performance among listed Indian companies.

On the contrary, Thammaraksa et al. (2024) emphasize the lack of significant statistical correlation in their sample between sustainability performance and SDG reporting. Moreover, Heras-Saizarbitoria et al. (2022) posit that SDG engagement by reporting companies does not go beyond being merely "superficial". In addition, Silva (2021) concludes that SDG reporting practices are fairly symbolic and not aimed at transformation.

The empirical papers cited above lead to an important result. Although the SDGs can easily be integrated into companies' reporting practices in theory, in practice, there are significant divergences. As discussed before, stakeholders should be able to compare corporations based on their SDG performance, using SDGs as a soft law instrument to assess their degree of responsibility. Significant divergences in SDG disclosures undermine the ultimate goal of ensuring comparative, concise, and complete information on corporations' SDG performance and of tilting them towards better sustainability performance. In addition, incoherent SDG reporting practices amplify "SDG-washing" risks.[26]

The underlying reason for these shortcomings lies in the absence of a recognized gatekeeping mechanism for SDG-related reporting. There are established verification institutions that assess sustainability characteristics and goals in other fields, such as the Science Based Targets Initiative, which certifies net-zero commitments. However, at the time of drafting of this contribution, no comparable assurance framework exists for SDGs. To enhance credibility and trust in SDG reporting, corporations should transition from self-assessment of their progress toward the goals to verification by independent third-party assurance providers. To address this gap, the International Organization for Standardization (**"ISO"**) is already developing a management system standard. Accordingly, ISO/UNDP DIS 53001 is set to become applicable in 2026 (Social Value International, 2025b) and will aim to acknowledge the integration of SDG-related measures into decision-making processes (ISO, 2025).

Another recent yet significant initiative aimed at strengthening third-party verification of SDG disclosures is the collaboration between the UN and Social Value International. Through the SDG Impact Standards Assurance Scheme, the UN seeks to establish a structured framework of standards and qualification criteria applicable to third-party assurance providers. Within this framework, assurance providers may obtain accreditation as SDG Impact Standards Accredited Practitioners upon completion of a designated training and certification program (Social Value International, 2025a). Should this mechanism, alongside the new ISO/UNDP DIS 53001 standard, achieve widespread adoption, SDG reporting practices would become more reliable, comparable, and, ultimately, more informative.

---

[26] For a recent example of SDG washing instance from Germany, see Bundesamt für Justiz (2025).

3 Fintech and SDGs: A Playing Field Where There Is More Room for "Doing Well by Doing Good"

3.1 The Concept of "Sustainability-Oriented Innovation"

Innovation has been a frequently visited concept in explaining production processes and competitiveness for a long time.[27] However, it has attracted increasing interest recently for several reasons. First of all, thanks to the inspiring works of the 2025 Nobel laureates, it is acknowledged that innovation and associated creative destruction are the main drivers of continuous economic growth (The Royal Swedish Academy of Sciences, 2025). Secondly, innovation can now target consumers more quickly and easily, thereby affecting people's everyday lives more significantly. Large language models are a prominent example of this. Thirdly, and as the most relevant to this contribution, innovation is increasingly being considered as an important instrument for sustainable development and green transition. (Acemoglu et al., 2016; Dechezleprêtre & Hémous, 2023; Voegtlin & Scherer, 2017). As established in the foregoing analysis, market participants have been increasingly demanding responsible business conduct, expanding the incentives to pursue sustainability-oriented innovation. (Klewitz & Hansen, 2014).

Within this context, sustainability-oriented innovation is pursued by corporations via two different channels: i) process innovation and ii) product innovation (OECD & Eurostat, 2018). Process innovation occurs, for example, through reduced resource consumption and effective control of non-product outputs such as waste (Klewitz & Hansen, 2014). Product innovation, on the other hand, relates to the enhancement of existing goods and services or the creation of entirely new ones (Klewitz & Hansen, 2014).[28] In both types, three dimensions of sustainable development are pursued, namely social, environmental, and economic (Kneipp et al., 2019).

As the preceding paragraph demonstrates, there is a strong conceptual proximity between sustainability-oriented innovation and the value dimension of business responsibility, as elaborated in Sect. 2.2. Within this context, sustainability-oriented innovation can lead to corporate value maximization while also contributing to non-shareholder stakeholder interests. Indeed, it was suggested more than fifteen years ago that sustainability should be considered a "new innovation frontier", where directors can reconcile the idea of being sustainable with maximizing company value ((Kneipp et al., 2019; Nidumolu et al., 2009). The extant literature sheds light on how corporations do this empirically.

First of all, as Aguilera-Caracuel and Ortiz-de-Mandojana (2013) explain comprehensively, firms seeking sustainability-oriented innovation can benefit from reduced

---

[27] According to the OECD, innovation is defined as follows: "*An innovation is a new or improved product or process (or combination thereof) that differs significantly from the unit's previous products or processes and that has been made available to potential users (product) or brought into use by the unit (process)*" (OECD & Eurostat, 2018).

[28] Although authors' primary focus in this paper is SMEs, their findings can reasonably be generalized to non-SME companies as well for the purposes of this contribution.

costs. Green innovative firms can restructure their production systems with effective environmental strategies and achieve a more organizationally efficient setting, through which they can reduce product life cycle costs and lower operational costs. Additionally, sustainability can drive novel ideas and visions, engendering fresh business opportunities (Hansen et al., 2009). Thereby, corporations can gain a competitive advantage in various areas through creation of enhanced value, reductions of costs, and acquisition or improvement of non-financial assets (Hermundsdottir & Aspelund, 2021). Ultimately, green innovation induces greater business pro-activity and organizational capabilities (Aguilera-Caracuel & Ortiz-de-Mandojana, 2013).[29] Aguilera-Caracuel and Ortiz-de-Mandojana (2013) further test these arguments empirically and find a statistically significant positive correlation between green innovation intensity, measured by green patent numbers, and enhanced financial performance, measured by return on assets at the firm level.[30]

Overall, sustainability-oriented innovation is a crucial concept that is conveniently situated within the explanations presented in the previous section. It enables corporations to act responsibly while serving the shareholders. However, this potential can be exploited by competent corporate directors dedicated to creating value. This is the case because sustainability-oriented innovation entails high risks regarding both economic success and sustainability impact (Hansen et al., 2009). In other words, it is challenging to predict ex ante the economic and sustainability impacts of a corporate action aimed at sustainability-oriented innovation, rendering its implementation difficult for corporate directors.

3.2 *Sustainability-Oriented Innovation Through Digitalization*

One prominent way to pursue sustainability-oriented innovation, as explained above, is to leverage digital tools and technology (Cisneros Chavira et al., 2023). To explain the interrelation between digital transformation and net-zero transition within the context of environmental goals, the concept of "twin transition" has been coined (World Economic Forum, 2022).

A recent OECD working paper provides a concise framework for the twin transition by highlighting its potential as well risks (Calvino et al., 2025). Accordingly,

[29] For a more recent paper conducting an extensive literature review and proposing a comprehensive list of benefits associated with green innovation, see Cisneros Chavira et al. (2023).

[30] Authors did not find a significant divergence among green innovative firms and non-green innovative firms. They propose that this might be the case because the financial performance impacts of green innovation require some time. Moreover, they posit that not all firms engaging in green innovation have the required conditions to achieve improvements in financial performance (Aguilera-Caracuel & Ortiz-de-Mandojana, 2013). For another paper exploiting surveys with Brazilian industrial companies, finding a significant correlation between sustainable innovation practices and various corporate performance measures, see Kneipp et al. (2019).

digitalization and net-zero transition are interacting with each other through the following five channels:

- **Energy use of capital:** This channel corresponds to the ratio of machines in corporate capital, thereby representing the aggregate energy use. The outcome of this interaction can be both positive and negative. In the best-case scenario, digitalization will lead to greater energy efficiency, resulting in less energy-intensive machines. In the worst-case scenario, due to increasing demand for computers, the use of machines will become more energy-intensive.
- **Capital intensity:** If machines gain a competitive advantage over workers in terms of output productivity, more machines and hardware will replace workers, resulting in increasing capital intensity and energy consumption.
- **Type of energy use:** Digitalization can support the productivity of clean technologies, contributing to environmental goals. For example, the battery performance of electric cars can be enhanced through machine learning-based optimization. This way, it can indirectly support electric vehicles over those powered by internal combustion engine vehicles.
- **Direction of innovation:** Digital-based innovation can serve research and development efforts in either clean or dirty technologies, depending on their potential to accommodate innovation.
- **Scale:** Digital-based innovation can amplify the aggregate demand in the economy, which would translate into higher energy consumption.

This brief summary shows that digital-based innovation is a double-edged sword for environmental goals. It holds huge potential that can be realized through these channels, as well as significant risks. One empirical insight acknowledging this tension relates to overall emissions of highly digital sectors and the information and communication technologies ("**ICT**") services sector. Accordingly, although there is an overall reduction in emissions from highly digital sectors, this reduction is not observed for ICT services, due to increasing energy consumption (Calvino et al., 2025; World Bank Group, 2024).

A similar relationship exists between digital-based innovation and social development, as they can interact in various ways. Three main channels are as follows (Nosratabadi et al., 2023):

- **Social inclusion:** Digital-based innovation can foster social inclusion by enabling people to connect with each other and access information and knowledge in ways that would not otherwise be possible. However, there is also a risk that societies will suffer from "digital divide". The digital divide occurs when there are unequal opportunities to benefit from the surplus generated by digital-based innovation (OECD, 2021).
- **Fundamental rights:** Digital-based innovation causes complications regarding various fundamental rights, such as those related to privacy, freedom of speech, and digital security. In this vein, digitization increases the risks of unauthorized use of personal data, online harassment and abuse, and exposure to harmful content. On the other hand, it can enhance safety by facilitating access to emergency

services, supporting innovative forms of community assistance and resilience, and fostering digital literacy and awareness of online threats (Nosratabadi et al., 2023).

- **Consumption and production:** As digitalization spreads across various fields of human life, interest is increasing in replacing traditional, resource-heavy physical goods and services with digital or "virtual" and intangible alternatives (Haskel & Westlake, 2018). Goods and services that are in virtual form are significantly less costly, and thus more affordable for low-income constituents. Additionally, using digital coordination platforms enables meeting demand with supply significantly easier and more efficiently, revolutionizing business practices. However, without proper governance, digitalization can result in the concentration of power through centralized control over data, heightening the potential for information dominance and large-scale manipulation, as well as systemic risks (Nakicenovic et al., 2019).

There are numerous other channels that digital-based innovations interact with social goals such as health, gender equality, food, smart cities, and other areas (Nakicenovic et al., 2019). Yet, it should be borne in mind that this interaction carries both significant potentials and risks.

3.3 Reflections in the Financial Sector: The Case of FinTech and SDGs

In the financial sector, technology-based innovation towards sustainability is spearheaded by so-called "fintech". Fintech leverages technological developments and disrupts the incumbent way of providing financial products and services. It enables new entrants into the sector and creates new opportunities to gain a competitive advantage for traditional financial institutions (Lee & Shin, 2018).

The novelties that fintech offers span the whole financial services market. In this vein, fintech startups, as the primary drivers of entrepreneurial innovation, have developed innovative business models in various fields, including payment services, wealth management, crowdfunding, lending, capital markets, and insurance (Lee & Shin, 2018). This is made possible thanks to technology developers who deliver the foundational infrastructure for digital innovation, including social media platforms, big data analytics, cloud services, artificial intelligence, smartphones, and mobile technologies (Lee & Shin, 2018), as well as blockchain technologies (Nakicenovic et al., 2019).

This relationship between fintech companies and technology developers benefits more than the parties involved and creates significant positive externalities. These benefits originate from the interaction between digitalization and sustainability-oriented innovation. Accordingly, the new business models offered by fintech are closely interrelated with sustainable development and SDGs in particular. Hasan et al. (2024) disentangle this relationship by conducting a comprehensive literature review and grouping it into three themes, namely (i) financial impact, (ii) economic

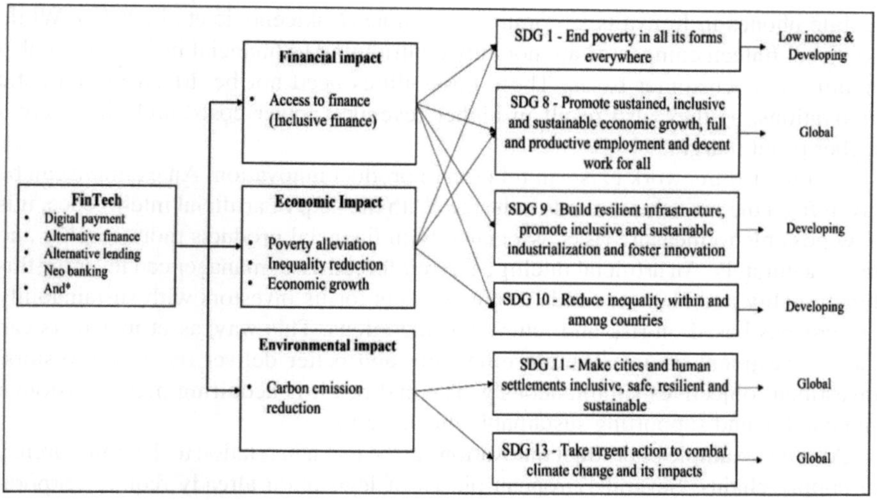

Fig. 2 Interaction between Fintech and specific SDGs (Hasan et al., 2024)

impact, and iii) environmental impact. The paper further conceptually maps the contribution of fintech to sustainable development onto specific SDGs, yielding the following chart[31] (Fig. 2).

The channels through which fintech can contribute to the SDGs have been examined on several occasions throughout this book. The key focus here is on how this interaction benefits both sustainable development and corporate value. Fintech firms hold substantial potential in that regard, as their business models are primarily based on digital innovation. By deploying innovative efforts towards sustainability, fintech companies can maximize their value and contribute to non-shareholder stakeholder interests simultaneously. This can be realized in both forms of process innovation and product innovation (Voegtlin & Scherer, 2017).

From a process innovation perspective, digitalization and technology transform business models in finance. This transformation enables the delivery of financial services and products in a more efficient and sustainable manner. For example, it is now possible to provide financial services with significantly less physical capital and, in turn, a substantially reduced carbon footprint (Offiong et al., 2025).[32] This translates into higher cost efficiency and better environmental performance at the same time. Similarly, by leveraging consumers' growing use of digital devices, fintech companies can now offer products and services that would otherwise be impossible, thereby contributing to financial inclusion. The provision of financial services via

[31] Please note that this chart is not conclusive, and other scholars explore the link between fintech and other SDGs such as SDG5 (Noreen, 2024), SDG7, SDG12, SDG14, SDG15, and SDG17 (Puschmann et al., 2020).

[32] For an empirical paper concluding that fintech significantly improves carbon emission efficiency, see Zhang et al. (2024).

mobile phones in Kenya is a prominent example (Nakicenovic et al., 2019). While doing so, fintech companies are not only contributing to financial inclusion but also creating new customer bases. These possibilities need not be driven by altruistic motivations, as they also result in higher revenues, lower costs, and ultimately a higher profit margin.

A similar framework exists in relation to product innovation. An example can be given from the investment funds industry. With the help of artificial intelligence, it is now possible to measure risks associated with financial products more quickly and more accurately. An artificial intelligence-reinforced asset manager can leverage this functionality and channel funds to green assets for its investors with sustainability preferences based on risk and return considerations. This way, asset managers can make investment decisions more efficiently and better deliver on their investors' investment objectives simultaneously. The end result is acquiring higher customer satisfaction and supporting sustainable finance efforts.

Another example of product innovation can be given in relation to the transparency of supply chains. Several European pieces of legislation already require corporations to identify the sustainability characteristics of the companies in their value chains (e.g., Corporate Sustainability Reporting Directive, Corporate Sustainability Due Diligence Directive). However, as indicated by a recent report, data availability is the primary challenge particularly for companies' scope three disclosures with respect to the environmental performance (Velázquez Martínez et al., 2025). The use of blockchain technologies has the potential to significantly alleviate these challenges and reduce the costs of obtaining this information, while also ensuring its trustworthiness.

These cases yield an important conclusion: There is ample room for fintech to serve sustainable development and financial performance simultaneously. This is possible because of the convenience of integrating sustainability-oriented innovation into fintech business models. As explained before, sustainability-oriented innovation, particularly when pursued through digitalization, has the inherent potential to support the financial performance of corporations. This is the case even when the primary motivation for utilizing innovation is not to serve sustainable development goals, but rather to reduce costs or increase operating margins. This convenience makes fintech a unique phenomenon that makes doing well by doing good relatively more possible.

One important implication of this conclusion pertains to the explanations under the Sect. 2.1. As discussed, corporate law rules require directors to prioritize shareholder wealth over other stakeholders, which exposes them to the challenge of dealing with trade-offs while engaging with responsible corporate actions. However, fintech is a peculiar playing field where serving sustainable development goals would also likely result in corporate value. This is crucial because the more aligned the sustainability goals and profit-maximizing opportunities for companies are, the wider the wiggle room for corporate directors. Such a relief has the potential to trickle down to institutional investors as well, who are themselves subject to fiduciary duties vis-à-vis their own investors. A recent survey of asset managers reveals that one of the main constraints on observing sustainability performances of portfolio companies by asset managers is concerns about their legal fiduciary duties, even for funds that market themselves as sustainable (Edmans et al., 2025). Such concerns can be relatively less significant within the context of fintech due to their peculiar case.

3.4 Being Responsible by Managing Risks

A final important aspect of business responsibility within the context of fintech is risk management. It is well established that regulating fintech is a very challenging task. The ultimate goal is relatively simple to establish: fostering economic development spurred by innovation while addressing risks to financial stability and consumer protection (Arner et al., 2017; Zetzsche et al., 2017). However, this is difficult to implement primarily because technology developers and, in turn, fintech companies consistently outpace regulators in developing innovative business models and identifying their risks, resulting in various inconsistent regulatory approaches across jurisdictions (Appaya & Gradstein, 2020; Arner et al., 2017). This misalignment became more pronounced as the use of artificial intelligence in the financial sector widened (Murinde et al., 2022).

Against this background, to avoid regulatory failure, regulators may prefer not to impose strict rules or to create regulatory sandboxes, which ultimately means more leeway for fintech companies in managing the risks their activities pose to other stakeholders. In this setting, the legal domain tides back, opening more room for business responsibility. Therefore, fintech companies that manage risks associated with their activities successfully and do not assume excessive risks can be deemed responsible. The most acute risks associated with fintech companies include cyber risks, data risks, consumer protection and capabilities risks, anti-money laundering and combating the financing of terrorism risks, financial stability, and disintermediation and excess volatility risks (Appaya & Gradstein, 2020). Managing such risks can be a more significant aspect of business responsibility, particularly in Europe, given the acceleration of policymakers' simplification and deregulation efforts following the Draghi report on EU competitiveness (European Commission, 2025).

4 Conclusion

Business responsibility is a relative concept that can adapt to the values and views prevalent in a society at a given time. This flexibility allows it to maintain its popularity as an area of debate that scholars across disciplines have been grappling with for a long time. Legal scholars have a special interest in this debate, not least because the corporate law rules prescribe how a director should govern a company and to what extent they provide legroom to implement responsible business practices. The investigation in the first subsection showed that the dominant regulatory approach across jurisdictions under corporate law rules is to require corporate directors to pursue shareholder value maximization in fulfilling their fiduciary duties, primarily because structural mechanisms to credibly commit directors to non-stakeholder interests are unavailable. Building on this finding, businesses are recognized as economic agents in society that respond to incentives, rather than moral agents that can pursue altruistic motivations. Therefore, responsible businesses are identified as corporations that go beyond complying with regulatory requirements to pursue shareholder value maximization by either sharing the wealth they generate with non-shareholder stakeholders or reducing the negative externalities they impose on others on a voluntary basis.

The next subsection focused on the capital asset pricing model in connection with responsible businesses and summarized why and how corporations engage in responsible practices. By introducing the enlightened shareholder value theory, it reconciled the notion of responsible business practices with shareholder value maximization, emphasizing both the potential for value creation and the difficulties of implementation.

The following subsection summarized the importance and functions of soft law in delivering on SDGs and suggested that SDGs serve as an important benchmark for industry, guiding corporations in their pursuit of responsible business practices. By referring to empirical research, it was found that SDGs are increasingly popular in corporate reporting practices. Yet, the reports seem far from informing investors in a standardized and comparable manner, undermining the purpose of disclosure in the first place. In an optimistic projection, it is suggested that once assurance services become mainstream practice, the identified shortcomings will be attenuated.

Later, the concept of sustainability-oriented innovation is introduced. It is explained as a potential driver of growth in the future that can generate corporate value for corporate shareholders while also contributing to the welfare of non-shareholder stakeholders. By situating fintech companies within this framework, the channels through which they can contribute to the SDGs, as well as the risks they pose, are illustrated.

Overall, the analysis in this contribution showed that business responsibility, sustainable development, and technological innovation are deeply interconnected and fintech represents a convenient convergence of these three dimensions within the financial sector. This convergence is likely to attract growing scholarly attention and to support the continued development of the fintech industry.

References

Acemoglu, D., Akcigit, U., Hanley, D., & Kerr, W. (2016). Transition to clean technology. *Journal of Political Economy, 124*(1), 52–104. https://doi.org/10.1086/684511

Aguilera-Caracuel, J., & Ortiz-de-Mandojana, N. (2013). Green innovation and financial performance: An institutional approach. *Organization & Environment, 26*(4), 365–385. https://doi.org/10.1177/1086026613507931

Appaya, M. S., & Gradstein, H. L. (2020). *How regulators respond to FinTech: Evaluating the different approaches—Sandboxes and beyond* (No. 148121). World Bank Group. http://documents.worldbank.org/curated/en/579101587660589857

Arena, M., Azzone, G., Ratti, S., Urbano, V. M., & Vecchio, G. (2023). Sustainable development goals and corporate reporting: An empirical investigation of the oil and gas industry. *Sustainable Development, 31*(1), 12–25. https://doi.org/10.1002/sd.2369

Arner, D. W., Zetzsche, D. A., Buckley, R. P., & Barberis, J. N. (2017). FinTech and RegTech: Enabling innovation while preserving financial stability. *Georgetown Journal of International Affairs, 18*(3), 47–58.

Arora-Jonsson, S. (2023). The sustainable development goals: A universalist promise for the future. *Futures, 146*, Article 103087. https://doi.org/10.1016/j.futures.2022.103087

Bainbridge, S. M. (2023). *The profit motive: Defending Shareholder value maximization* (1st ed.). Cambridge University Press. https://doi.org/10.1017/9781009025799

Barker, S., Williams, C., & Cooper, A. (2021). *Fiduciary duties and climate change in the United States*. Commonwealth and Climate Law Initiative.

Bebchuk, L. A., & Tallarita, R. (2020). The illusory promise of stakeholder governance. *Cornell Law Review, 106*, 91–178. https://doi.org/10.2139/ssrn.3544978

Bénabou, R., & Tirole, J. (2010). Individual and corporate social responsibility. *Economica, 77*(305), 1–19. https://doi.org/10.1111/j.1468-0335.2009.00843.x

Berchicci, L., & King, A. A. (2022). Corporate sustainability: A model uncertainty analysis of materiality. *Journal of Financial Reporting, 7*(2), 43–74. https://doi.org/10.2308/JFR-2021-022

Beretta, V., Demartini, M. C., & Trucco, S. (2025). SDG reporting and sustainability performance: Insights from Italian companies. *Social Responsibility Journal, 21*(8), 1657–1685. https://doi.org/10.1108/SRJ-07-2024-0492

Biermann, F., Kanie, N., & Kim, R. E. (2017). Global governance by goal-setting: The novel approach of the UN sustainable development goals. *Current Opinion in Environmental Sustainability, 26–27*, 26–31. https://doi.org/10.1016/j.cosust.2017.01.010

Binder, J.-H. (2024). Leading wherever they want? CSR, ESG, and directors' duties. In J.-H. Binder, K. Hopt, & T. Kuntz (Eds.), *Corporate purpose, CSR, and ESG* (1st ed., pp. 217–246). Oxford University Press. https://doi.org/10.1093/oso/9780198912576.003.0011

Blair, M. M., & Stout, L. A. (1999). A team production theory of corporate law. *Virginia Law Review, 85*(2), 247–328. https://doi.org/10.2307/1073662

Bundesamt für Justiz. (2025). *Injunction against Generali Investments Holding SpA, German branch.* https://www.bundesjustizamt.de/DE/Themen/Verbraucherrechte/VerbandsklageregisterMusterfeststellungsklagenregister/Verbandsklagenregister/Unterlassungsklagen/Klagen/2025/187/UKlag_187_2025_node.html#doc241910bodyText4

Business & Human Rights Resource Centre. (2025, June 26). *Spotify faces boycott over CEO's $700m investment in AI military defence startup.* Business & Human Rights Resource Centre. https://www.business-humanrights.org/en/latest-news/spotify-faces-boycott-over-ceos-700m-investment-in-ai-military-defence-startup/

Calvino, F., Dechezleprêtre, A., & Haerle, D. (2025). *How can the digital transformation affect the net-zero transition?: A conceptual framework for the "twin" transition* (OECD Science, Technology and Industry Working Papers) [OECD Science, Technology and Industry Working Papers]. https://doi.org/10.1787/4ee71f6d-en

Cashore, B. (2002). Legitimacy and the privatization of environmental governance: How non–state market–driven (NSMD) governance systems gain rule–making authority. *Governance, 15.*

Christensen, H. B., Hail, L., & Leuz, C. (2021). Mandatory CSR and sustainability reporting: Economic analysis and literature review. *Review of Accounting Studies, 26*(3), 1176–1248. https://doi.org/10.1007/s11142-021-09609-5

Cisneros Chavira, P., Shamsuzzoha, A., Kuusniemi, H., & Jovanovski, B. (2023). Defining green innovation, its impact, and cycle—A literature analysis. *Cleaner Engineering and Technology, 17*, Article 100693. https://doi.org/10.1016/j.clet.2023.100693

Coglianese, C., & Kagan, R. A. (Eds.). (2007). *Regulation and regulatory processes.* Ashgate.

Coglianese, C., & Mendelson, E. (2010). Meta-regulation and self-regulation. In C. Coglianese & E. Mendelson, *The Oxford handbook of regulation* (pp. 145–168). Oxford University Press. https://doi.org/10.1093/oxfordhb/9780199560219.003.0008

Condon, M. (2025). "Green" corporate governance. In J. N. Gordon & W.-G. Ringe (Eds.), *The Oxford handbook of corporate law and governance* (2nd ed.). Oxford University Press. https://doi.org/10.1093/oxfordhb/9780192888006.013.0033

Cornell, B., & Damodaran, A. (2020). Valuing ESG: Doing good or sounding good? *The Journal of Impact and ESG Investing, 1*(1), 76–93. https://doi.org/10.3905/jesg.2020.1.1.076

Dechezleprêtre, A., & Hémous, D. (2023). Chapter 16 directed technical change and environmental economics. In U. Akcigit & J. Van Reenen (Eds.), *The economics of creative destruction* (pp. 435–472). Harvard University Press. https://doi.org/10.4159/9780674293052-018

Edmans, A. (2011). Does the stock market fully value intangibles? Employee satisfaction and equity prices. *Journal of Financial Economics, 101*(3), 621–640. https://doi.org/10.1016/j.jfineco.2011.03.021

Edmans, A. (2022). *Grow the pie: How great companies deliver both purpose and profit* (Revised first paperback edition). Cambridge University Press.

Edmans, A. (2023a). The end of ESG. *Financial Management, 52*(1), 3–17. https://doi.org/10.1111/fima.12413

Edmans, A. (2023b, October 1). What social responsibilities should companies have? A new approach. *Wall Street Journal*. https://www.wsj.com/business/c-suite/social-responsibility-companies-96b0786c

Edmans, A., Gosling, T., & Jenter, D. (2025). *Sustainable investing in practice: objectives, beliefs, and limits to impact*. HKU Jockey Club Enterprise Sustainability Global Research Institute, European Corporate Governance Institute, Finance Working Paper.

Elalfy, A., Weber, O., & Geobey, S. (2021). The sustainable development goals (SDGs): A rising tide lifts all boats? Global reporting implications in a post SDGs world. *Journal of Applied Accounting Research, 22*(3), 557–575. https://doi.org/10.1108/JAAR-06-2020-0116

Enriques, L., & Romano, A. (2022). Rewiring corporate law for an interconnected world. *Arizona Law Review, 64*(1), 51–87.

Erin, O. A., & Bamigboye, O. A. (2022). Evaluation and analysis of SDG reporting: Evidence from Africa. *Journal of Accounting & Organizational Change, 18*(3), 369–396. https://doi.org/10.1108/JAOC-02-2020-0025

European Commission. (2011). *Communication from the Commission to the European Parliament, The Council, The European Economic and Social Committee and the Committee of the Regions*. European Commission.

European Commission. (2025). *The Draghi report on EU competitiveness*. The Draghi Report on EU Competitiveness. https://commission.europa.eu/topics/eu-competitiveness/draghi-report_en

European Commission & EY. (2020). *Study on directors' duties and sustainable corporate governance: Final report*. Publications Office. https://data.europa.eu/doi/10.2838/472901

Falck, O., & Heblich, S. (2007). Corporate social responsibility: Doing well by doing good. *Business Horizons, 50*(3), 247–254. https://doi.org/10.1016/j.bushor.2006.12.002

Ferrarini, G. (2025). Firm value versus social value: Dealing with the Trade-Offs. In K. Alexander, M. Gargantini, & M. Siri (Eds.), *The Cambridge handbook of EU sustainable finance* (1st ed., pp. 51–79). Cambridge University Press. https://doi.org/10.1017/9781009483971.006

Flammer, C. (2015). Does corporate social responsibility lead to superior financial performance? A regression discontinuity approach. *Management Science, 61*(11), 2549–2568. https://doi.org/10.1287/mnsc.2014.2038

Flammer, C., & Ioannou, I. (2021). Strategic management during the financial crisis: How firms adjust their strategic investments in response to credit market disruptions. *Strategic Management Journal, 42*(7), 1275–1298. https://doi.org/10.1002/smj.3265

Flammer, C., & Kacperczyk, A. (2016). The impact of stakeholder orientation on innovation: Evidence from a natural experiment. *Management Science, 62*(7), 1982–2001. https://doi.org/10.1287/mnsc.2015.2229

Friedman, M. (1970, September 13). A Friedman doctrine—The social responsibility of business is to increase its profits. *The New York Times*. https://www.nytimes.com/1970/09/13/archives/a-friedman-doctrine-the-social-responsibility-of-business-is-to.html

Gadinis, S., & Fregoni, S. (2025). Beyond the Brussels effect: The surprising rise of the international sustainability standards board. *SSRN Electronic Journal*.

Galeazzo, A., Miandar, T., & Carraro, M. (2024). SDGs in corporate responsibility reporting: A longitudinal investigation of institutional determinants and financial performance. *Journal of Management and Governance, 28*(1), 113–136. https://doi.org/10.1007/s10997-023-09671-y

Garst, J., Maas, K., & Suijs, J. (2022). Materiality assessment is an art, not a science: Selecting ESG topics for sustainability reports. *California Management Review, 65*(1), 64–90. https://doi.org/10.1177/00081256221120692

Gatti, M. (2025). *Corporate power and the politics of change* (1st ed.). Cambridge University Press. https://doi.org/10.1017/9781009704465

Gibson, C., Sawyer, F., & Siegel, J. (2025). *2025 statehouse report*. Pleiades Strategy. https://perma.cc/RP62-ZPWH

Gordon, J. N. (2022). Systematic stewardship. *Journal of Corporation Law, 47*, 627–673. https://doi.org/10.2139/ssrn.3782814

Green, J. F. (2010). Private standards in the climate regime: The greenhouse gas protocol. *Business and Politics, 12*(3), 1–37. https://doi.org/10.2202/1469-3569.1318

Hahn, T., Figge, F., Pinkse, J., & Preuss, L. (2018). A paradox perspective on corporate sustainability: Descriptive, instrumental, and normative aspects. *Journal of Business Ethics, 148*(2), 235–248. https://doi.org/10.1007/s10551-017-3587-2

Hansen, E. G., Grosse-Dunker, F., & Reichwald, R. (2009). Sustainability innovation cube—A framework to evaluate sustainability-oriented innovations. *International Journal of Innovation Management, 13*(04), 683–713. https://doi.org/10.1142/S1363919609002479

Hart, O., & Zingales, L. (2017). Companies should maximize shareholder welfare not market value. *Journal of Law, Finance, and Accounting, 2*(2), 247–275. https://doi.org/10.1561/108.00000022

Hart, O., & Zingales, L. (2024). The new corporate governance. In L. Enriques & G. Strampelli (Eds.), *Board-shareholder dialogue* (1st ed., pp. 24–45). Cambridge University Press. https://doi.org/10.1017/9781009360746.003

Hart, S. L. (1995). A natural-resource-based view of the firm. *The Academy of Management Review, 20*(4), 986–1014. https://doi.org/10.2307/258963

Hasan, M., Hoque, A., Abedin, M. Z., & Gasbarro, D. (2024). FinTech and sustainable development: A systematic thematic analysis using human- and machine-generated processing. *International Review of Financial Analysis, 95*, Article 103473. https://doi.org/10.1016/j.irfa.2024.103473

Haskel, J. E., & Westlake, S. (2018). *Capitalism without capital: The rise of the intangible economy*. Princeton university press.

Heras-Saizarbitoria, I., Urbieta, L., & Boiral, O. (2022). Organizations' engagement with sustainable development goals: From cherry-picking to SDG-washing? *Corporate Social Responsibility and Environmental Management, 29*(2), 316–328. https://doi.org/10.1002/csr.2202

Hermundsdottir, F., & Aspelund, A. (2021). Sustainability innovations and firm competitiveness: A review. *Journal of Cleaner Production, 280*, Article 124715. https://doi.org/10.1016/j.jclepro.2020.124715

Hopt, K. J. (2024). Corporate purpose and stakeholder value: Historical, economic, and comparative law remarks on the current debate, legislative options, and enforcement problems. In J.-H. Binder, K. Hopt, & T. Kuntz (Eds.), *Corporate Purpose, CSR, and ESG* (1st ed., pp. 17–54). Oxford University Press. https://doi.org/10.1093/oso/9780198912576.003.0002

Hummel, K., & Szekely, M. (2022). Disclosure on the sustainable development goals—Evidence from Europe. *Accounting in Europe, 19*(1), 152–189. https://doi.org/10.1080/17449480.2021.1894347

Hwang, C., & Lund, D. S. (2025). Purpose and Nonprofit Enterprise. *Columbia Law Review, 125*, 2267–2330. https://doi.org/10.2139/ssrn.5061739

ISO. (2025). *ISO/UNDP DIS 53001*. ISO. https://www.iso.org/standard/86672.html

Jensen, M. C. (2010). Value maximization, stakeholder theory, and the corporate objective function. *Journal of Applied Corporate Finance, 22*(1), 32–42. https://doi.org/10.1111/j.1745-6622.2010.00259.x

Khan, M., Serafeim, G., & Yoon, A. (2016). Corporate sustainability: First evidence on materiality. *The Accounting Review, 91*(6), 1697–1724. https://doi.org/10.2308/accr-51383

Kitzmueller, M., & Shimshack, J. (2012). Economic Perspectives on Corporate Social Responsibility. *Journal of Economic Literature, 50*(1), 51–84. https://doi.org/10.1257/jel.50.1.51

Klewitz, J., & Hansen, E. G. (2014). Sustainability-oriented innovation of SMEs: A systematic review. *Journal of Cleaner Production, 65*, 57–75. https://doi.org/10.1016/j.jclepro.2013.07.017

Kneipp, J. M., Gomes, C. M., Bichueti, R. S., Frizzo, K., & Perlin, A. P. (2019). Sustainable innovation practices and their relationship with the performance of industrial companies. *Revista De Gestão, 26*(2), 94–111. https://doi.org/10.1108/REGE-01-2018-0005

Koehler, G. (2016). Assessing the SDGs from the standpoint of eco-social policy: Using the SDGs subversively. *Journal of International and Comparative Social Policy, 32*(2), 149–164. https://doi.org/10.1080/21699763.2016.1198715

Kuntz, T. (2024). How ESG is weakening the business judgement rule. In T. Kuntz (Ed.), *Research handbook on environmental, social and corporate governance* (pp. 67–94). Edward Elgar Publishing. https://doi.org/10.4337/9781802202533.00011

Lee, I., & Shin, Y. J. (2018). Fintech: Ecosystem, business models, investment decisions, and challenges. *Business Horizons, 61*(1), 35–46. https://doi.org/10.1016/j.bushor.2017.09.003

Liang, H., Ng, L., & Yoon, A. (2025). Editorial: What have we learned about green and climate finance? *The British Accounting Review, 57*(5), Article 101749. https://doi.org/10.1016/j.bar.2025.101749

Lipton, A. (2025). The legitimation of shareholder primacy. *The Journal of Corporation Law, 83*, 83–125. https://doi.org/10.2139/ssrn.5120765

Long, G. (2015). The idea of universality in the sustainable development goals. *Ethics & International Affairs, 29*(2), 203–222. https://doi.org/10.1017/S0892679415000076

Lukács, B., Tóth, Á., & Molnár, P. (2025). Measuring corporate compliance with the SDGs Based on the GRI's ESG reporting methodology. *Journal of Sustainability Research, 7*(1). https://doi.org/10.20900/jsr20250010

Lund, D. S. (2021). Enlightened shareholder value, stakeholderism, and the quest for managerial accountability. In E. Pollman & R. B. Thompson (Eds.), *Research handbook on corporate purpose and personhood*. Edward Elgar Publishing. https://doi.org/10.4337/9781789902914.00012

Lund, D. S. (2025). Toward a dynamic view of corporate purpose. *Minnesota Law Review, 109*, 2089–2161. https://doi.org/10.2139/ssrn.4665040

Lund, D. S., & Pollman, E. (2021). The corporate governance machine. *Columbia Law Review, 121*. https://doi.org/10.2139/ssrn.3775846

Masulis, R. W., & Reza, S. W. (2015). Agency problems of corporate philanthropy. *The Review of Financial Studies, 28*(2), 592–636. https://doi.org/10.1093/rfs/hhu082

Mayer, C. (2024). *Success, law and ESG*. SSRN. https://doi.org/10.2139/ssrn.4939931

Murinde, V., Rizopoulos, E., & Zachariadis, M. (2022). The impact of the FinTech revolution on the future of banking: Opportunities and risks. *International Review of Financial Analysis, 81*, Article 102103. https://doi.org/10.1016/j.irfa.2022.102103

Nakicenovic, N., Messner, D., Zimm, C., Clarke, G., Rockström, J., Aguiar, A. P. D., Boza-Kiss, B., Campagnolo, L., Chabay, I., Collste, D., Comolli, L., Gomez-Echeverri, L., Goujon, A., Grubler, A., Jung, R., Kamei, M., Kamiya, G., Kriegler, E., Kuhn, M., … Zusman, E. (2019). *The digital revolution and sustainable development: Opportunities and challenges*. Report prepared by the World in 2050 initiative. https://doi.org/10.22022/TNT/05-2019.15913

Nidumolu, R., Prahalad, C. K., & Rangaswami, M. R. (2009). Why sustainability is now the key driver of innovation. *Harvard Business Review, 87*(9), 56–64.

Nolan, J. (2013). The corporate responsibility to respect human rights: Soft law or not law? In S. Deva & D. Bilchitz (Eds.), *Human rights obligations of business* (1st ed., pp. 138–161). Cambridge University Press. https://doi.org/10.1017/CBO9781139568333.010

Noreen, U. (2024). Mapping of FinTech ecosystem to sustainable development goals (SDGs): Saudi Arabia's landscape. *Sustainability, 16*(21), 9362. https://doi.org/10.3390/su16219362

Nosratabadi, S., Atobishi, T., & Hegedűs, S. (2023). Social sustainability of digital transformation: Empirical evidence from EU-27 countries. *Administrative Sciences, 13*(5), 126. https://doi.org/10.3390/admsci13050126

OECD. (2021). Bridging digital divides in G20 countries. *OECD Publishing*. https://doi.org/10.1787/35c1d850-en

OECD. (2025). *Responsible business conduct*. https://www.oecd.org/en/topics/responsible-business-conduct.html

OECD & Eurostat. (2018). *Oslo manual 2018: Guidelines for collecting, reporting and using data on innovation* (4th ed.). OECD Publishing. https://doi.org/10.1787/9789264304604-en

Offiong, U. P., Szopik-Depczyńska, K., & Ioppolo, G. (2025). FinTech innovations for sustainable development: A comprehensive literature review and future directions. *Sustainable Development*, 1–13. https://doi.org/10.1002/sd.70068

Osman, A. I., Chen, L., Yang, M., Msigwa, G., Farghali, M., Fawzy, S., Rooney, D. W., & Yap, P.-S. (2023). Cost, environmental impact, and resilience of renewable energy under a changing climate: A review. *Environmental Chemistry Letters, 21*(2), 741–764. https://doi.org/10.1007/s10311-022-01532-8

Osterloh-Konrad, C. (2024). ESG and the ethical dimension. In J.-H. Binder, K. Hopt, & T. Kuntz (Eds.), *Corporate purpose, CSR, and ESG* (1st ed., pp. 161–172). Oxford University Press. https://doi.org/10.1093/oso/9780198912576.003.0008

Puschmann, T., Hoffmann, C. H., & Khmarskyi, V. (2020). How Green FinTech can alleviate the impact of climate change—The case of Switzerland. *Sustainability, 12*(24), 10691. https://doi.org/10.3390/su122410691

Roe, M., Spamann, H., Fried, J., & Wang, C. (2021). The sustainable corporate governance initiative in Europe. *Yale Journal on Regulation Bulletin, 38*.

Rosati, F., & Faria, L. G. D. (2019). Addressing the SDGs in sustainability reports: The relationship with institutional factors. *Journal of Cleaner Production, 215*, 1312–1326. https://doi.org/10.1016/j.jclepro.2018.12.107

Schanzenbach, M. M., & Sitkoff, R. H. (2020). Reconciling fiduciary duty and social conscience. *Stanford Law Review, 72*, 381–454.

SDG Impact. (2023). *About the SDG impact standards*. UNDP.

Shaffer, G. C., & Pollack, M. A. (2010). Hard vs. soft law: Alternatives, complements, and antagonists in international governance. *Minnesota Law Review, 94*, 706–799.

Sharma, R., Chawla, S., Dagar, V., Kagzi, M., & Rao, A. (2025). SDG adoption and firm risk: The impact of ESG performance, investor confidence, and agency cost. *International Review of Economics & Finance, 101*, Article 104205. https://doi.org/10.1016/j.iref.2025.104205

Silva, S. (2021). Corporate contributions to the sustainable development goals: An empirical analysis informed by legitimacy theory. *Journal of Cleaner Production, 292*, Article 125962. https://doi.org/10.1016/j.jclepro.2021.125962

Snapes, L. (2025, September 18). Massive attack remove music from spotify to protest against CEO Daniel Ek's investment in AI military. *The Guardian*. https://www.theguardian.com/music/2025/sep/18/massive-attack-remove-music-from-spotify-to-protest-ceo-daniel-eks-investment-in-ai-military

Social Value International. (2025a). *SDG Impact Standards Assurance Scheme*. Social Value International. https://www.socialvalueint.org/sdg-assurance-scheme

Social Value International. (2025b, October 8). *ISO 53001/2 is coming in 2026, but here's how to get ahead now*. Social Value International. https://www.socialvalueint.org/blog/coming-soon-iso-53001-but-you-can-get-started-todaynbsp

Stout, L. A. (2012). New thinking on "Shareholder primacy." *Accounting, Economics, and Law, 2*(2).

Thammaraksa, C., Gebara, C. H., Hauschild, M. Z., Pontoppidan, C. A., & Laurent, A. (2024). Business reporting of SUSTAINABLE DEVELOPMENT GOALS: Global trends and implications. *Business Strategy and the Environment, 33*(6), 5445–5462. https://doi.org/10.1002/bse.3760

The Economist. (2019). *What companies are for*. https://www.economist.com/leaders/2019/08/22/what-companies-are-for

The Royal Swedish Academy of Sciences. (2025). *The prize in economic sciences 2025—Popular science background*.

UN Global Compact. (2025a). *SDG action manager | UN global compact*. https://unglobalcompact.org/take-action/sdg-action-manager

UN Global Compact. (2025b). *SDG integration-UN global compact*. https://unglobalcompact.org/what-is-gc/our-work/sdg-integration

UNIDO. (2025). *What is CSR?* UNIDO. https://www.unido.org/our-focus/advancing-economic-competitiveness/competitive-trade-capacities-and-corporate-responsibility/corporate-social-responsibility-market-integration/what-csr

Velázquez Martínez, J. C., Rajagopalan, S., Arnold, V., & Quiñones, C. A. M. (2025). *State of supply chain sustainability 2025* (MIT Center for Transportation & Logistics and Council of Supply Chain Management Professionals.). https://sustainable.mit.edu/sscs-report/

Voegtlin, C., & Scherer, A. G. (2017). Responsible innovation and the innovation of responsibility: Governing sustainable development in a globalized world. *Journal of Business Ethics, 143*(2), 227–243. https://doi.org/10.1007/s10551-015-2769-z

World Bank Group. (2024). Measuring the emissions and energy footprint of the ICT sector: Implications for climate action. *The World Bank*. https://doi.org/10.1596/41238

World Economic Forum. (2022, October 26). *What is the 'twin transition'—And why is it key to sustainable growth?* World Economic Forum. https://www.weforum.org/stories/2022/10/twin-transition-playbook-3-phases-to-accelerate-sustainable-digitization/

Zetzsche, D. A., Buckley, R. P., Arner, D. W., & Barberis, J. N. (2017). Regulating a revolution: From regulatory sandboxes to smart regulation. *Fordham Journal of Corporate & Financial Law, 23*(1). https://doi.org/10.2139/ssrn.3018534

Zhang, Y., Chen, M., Zhong, S., & Liu, M. (2024). Fintech's role in carbon emission efficiency: Dynamic spatial analysis. *Scientific Reports, 14*(1), 23941. https://doi.org/10.1038/s41598-024-74834-2

State Obligations Under International Law, Role of Fintech in Achieving the SDGs, and Regulations Thereof

Hilal Koçali

1 Introduction

The Sustainable Development Goals (SDGs) addresses critical global challenges which have environmental, social and economic aspects as the achievement of SDGs requires a significant amount of financial undertaking, this emphasizes the importance of the financial sector, and therefore financial technologies (fintech).

Understanding the role of fintech in the context of the SDGs depends primarily on which obligations exist in this regard under international law. Indeed, once these obligations are understood, it will be possible to analyze what is being done in practice and whether it is sufficient. In order to achieve these obligations, regulatory measures are taken under domestic law, which also affects the finance sector and, consequently fintech strongly.

The regulatory activities directly impact the development and production of financial technologies. Likewise, the sustainable growth of financial technologies would not be feasible in countries with ill-conceived regulations. Indeed, regulatory stances of states could positively or negatively affect the development and sustainability of fintech, and this is significant given fintech's role in achieving the SDGs.

2 Definition and Sources of State Obligations

Oxford Classical Dictionary defines the obligation as "'a legal tie which binds us to the necessity of making some performance in accordance with the laws of our state'" (Zimmermann, 2016). Although state obligations under international law is a topic which is understudied (Bennouna, 2017; Demaria, 2021; Verhoeven, 2002) state

H. Koçali (✉)
Global Best Practice Group, Istanbul, Turkey
e-mail: Kocali@globalbestpracticegroup.org

B. Ş. Şahin and B. C. Cantürk (eds.), *Fintech and Sustainable Development*, Sustainable Finance, https://doi.org/10.1007/978-3-032-31020-0_7

obligations in general refer the commitment by states to be bound by an international law instrument (Lukashuk, 1989; Syed & Ikra, 2023). Among the many instruments of international law which state obligations are derived from (V.D. Degan, 1997; Thirlway, 2019), treaties and court judgments will be focused on this chapter within the context of the SDGs.

According to the International Court of Justice (ICJ), sustainable development is about the need to pursue economic development while safeguarding environmental resources (International Court of Justice, 1997). Within this understanding the Sustainable Development Goals (SDGs) was created to operationalize the concept of sustainable development under international cooperation (Hák et al., 2016; Sachs, 2015). SDGs refers to ongoing significant matters such as human rights, protection of environment, eradication of poverty and the aims thereof (Guiry, 2024a; United Nations Department of Public Information, 2017). In total, it consist of 17 goals and 169 targets across various issues (Katila et al., 2019). Although the SDGs are considered a framework rather than a legally binding instrument, many of them are derived from legally binding obligations, such as international treaties and customary international law (Kim, 2016). This makes certain aspects of SDGs, which were derived from legally binding obligations, are state obligations as well, and States are expected to use their governance to oblige as such (Irawan et al., 2025).

3 Environmental Dimension of Sustainable Development Goals

3.1 Legally Binding Instruments on Environmental Protection

Sustainability has three pillars: environmental, economic, and social (Purvis et al., 2019). SDGs which are related to the environment are SDG 13 (climate change), 14 (life below water), and 15 (life on land). Although SDGs are considered a framework rather than an international obligation, there are numerous legally binding instruments regarding the environmental protection. According to the International Court of Justice (ICJ), international environmental protection is governed by a number of binding legal instruments, among which the United Nations Framework Convention on Climate Change, and the Paris Agreement occupies a central position (International Court of Justice, 2025). Along with the multilateral environmental agreements addressing oceans, biodiversity, atmospheric protection, and land degradation, they together form the legal framework of the states' environmental obligations under international law.

3.2 The Paris Agreement: A Milestone in Environmental Obligations

Among others, the Paris Agreement is considered a milestone with respect to the environmental protection (Chatjuthamard et al., 2025). Indeed, under SDG 13 it is stated that the UN Framework Convention on Climate Change (UNFCCC), to which the Paris Agreement is a part of, is fundamental platform to tackle the climate change issue (United Nations General Assembly, 2015). The Paris Agreement was adopted at the UN Climate Change Conference (COP21) in Paris on December 12, 2015, and entered into force on November 4, 2016, the 30th day after at least 55 states that are responsible for 55% of total greenhouse gas emissions deposited their instruments of ratification (United Nations, 2015). The Paris Agreement is significant from a financial perspective as well, since it is a legally binding instrument and it dictates the obligation of "[making] finance flows consistent with a pathway toward low greenhouse gas emissions and climate-resilient development" (United Nations, 2015, p. 2, Article 2.1 (c)). This consists of an international obligation for the signatory states to utilize finance in order to achieve environment related goals.

Paris Agreement also includes an obligation for the state parties to take necessary actions to limit global warming to substantially below 2 °C above pre-industrial levels and to actively pursue measures aimed at capping temperature increases at 1.5 °C, given the significantly lower risks and adverse effects associated with that threshold (United Nations, 2015).

Therefore, the Paris Agreement dictates goal-specific obligations on the states who ratified it, which constitutes a state obligation under international law.

3.3 Nationally Determined Contributions (NDCs) and Financial Commitments

Responsibilities under the Agreement are monitored with a periodical submission mechanism, where in each five years signatory parties must submit an updated climate action plan—also known as Nationally Determined Contribution (NDC) (Dodson et al., 2022). Such periodic monitoring mechanisms are not unfamiliar to the UN, as member states periodically submit a report to undergo review on its human rights-related matters for Universal Periodic Review (UPR) since 2008 (Gayatri Patel, 2016).

Under the NDC submitted by Spain on behalf of the European Union and its member states, it is stated that implementation of Article 2.1(c) of the Paris Agreement is supported through regulations such as the EU Sustainable Finance Framework (European Union, 2023). Despite being significantly longer, Türkiye's NDC similarly provides the framework for sustainable finance without referring to the role of fintech in fulfilling obligations under the Paris Agreement (Republic of Türkiye, 2023).

The omission of fintech from NDCs is significant and demonstrates that the role of fintech in achieving the SDGs is yet to be recognized entirely. This could also stem from the lack of academic research on the relationship between fintech and carbon emission (Wu, 2024).

3.4 International Court Judgments Clarifying State Obligations

Alongside international treaties, one of the sources of international law are international court judgments. Although international law does not operate on a precedent basis, international courts' added value to the legal is undeniable (Tams, 2018). Despite the existence of numerous legally binding international texts relating to environmental protection, as referred to above, the case law of international courts on the subject is limited. That is mainly due to the difficulty of establishing jurisdiction on the issue, as facts are complex for an international court to examine (Zarei et al., 2015).

Nevertheless, this does not mean there is no case law which clarifies scope of state duties in addressing climate change and environmental protection. Recent judgment of the European Court of Human Rights (ECtHR) on the case of Verein Klimaseniorinnen Schweiz and others v. Switzerland constitutes the first judgment from an international court finding a violation against a state regarding the failure of mitigation of the negative impacts of climate change (Savaresi, 2025).

The case, which turned into a discussion topic also from a procedural point of view, such as *locus standi* (Bernard & Rugna, 2024), is brought before the ECtHR by four women and a Swiss association, whose members are elderly women, regarding the adverse effects of climate change on their health. The applicants claimed that Switzerland's actions to address these effects were insufficient and thus amounted a breach of their rights under the European Convention on Human Rights (ECHR). Although the Convention does not include climate change-specific obligations, ECtHR found jurisdiction on the matter and held that, as Swiss domestic regulatory framework was not developed and implemented adequately, Article 8 (right to family and private life) of the ECHR had been breached (European Court of Human Rights, 2024). The reasoning of the judgment supports the notion that climate inaction may constitute a human rights violation, hence reinforcing the legal framework of SDG commitments (European Court of Human Rights, 2024).

For the purposes of this chapter, it is sufficient to point that ECtHR judgments are only legally binding for respondent states and for other member states, they are authoritative because the Convention is a living instrument, and it is expected from member states to align their domestic law with the case law of the Court (Arnardóttir, 2017). Therefore, ECtHR judgments are legally binding at a certain level, and it is a state obligation for the 46 member states, including all EU member states and Türkiye, to abide by them (Grynchak et al., 2022). In simple terms this means that state obligations relating to the SDGs are also evolving through jurisprudence, and this is one of the many factors that will impact related regulatory-making.

As sources of international law explained above are legally binding for certain states, it could thus be stated that fulfilment of SDGs related to environmental protection is a state obligation for those, and states are expected to take action to fulfil these obligations. As a strong link between fintech and environmental protection has already been established, it is important to analyze what is being done by states to uphold these obligations, and achieve the SDGs through fintech, alongside other means.

4 The Socio-Economic Dimension of Sustainable Development Goals

Social rights ensures that the individuals can access the essential resources and conditions required for human dignity and well-being (Philipp, 2024), and socio-economic rights are considered an indispensable part of human rights (Davis, 2024).

As environmental obligations represent one pillar of sustainable development, the socio-economic component is grounded in economic, social, and cultural rights. Such rights are enshrined in international law in treaties and as well as legal frameworks.

4.1 The International Covenant on Economic, Social, and Cultural Rights

The International Covenant on Economic, Social and Cultural Rights (ICESCR), adopted in 1966, imposes legal obligations on a range of issues, including right to work, right to gain one's living through freely chosen work, right to social security and right to an adequate standard of living (United Nations, 1966), which are directly linked to SDG 1 (no poverty) and SDG 8 (economic growth) enshrined under SDGs. The Covenant is considered a landmark human rights system within the UN (Saul et al., 2014). Out of the 193 UN member states, 173 ratified the agreement. The large number of participants in the treaty confirms its importance. It also demonstrates that a significant number of UN member states are bound by the legal obligations set out therein.

4.2 The European Social Charter

The European Social Charter (ESC), similar to the ECHR, was opened for signature within the Council of Europe in 1961. The Revised European Social Charter was adopted in 1996, and both texts are currently in force. The ESC was drafted in response to the need for a text on social and economic rights, alongside the ECHR, which protects civil and political rights. Similar to the ICESCR, it addresses collective rather than individual rights (David Harris, 2021). In total, 42 Council of Europe member states out of 46 ratified the ESC, and only 16 accepted the collective complaint procedure (Council of Europe, 2025).

Article 30 of the ESC is significant as it obliges signatory states to "to take measures within the framework of an overall and co-ordinated approach to promote the effective access of persons who live or risk living in a situation of social exclusion or poverty, as well as their families, to, in particular, employment, housing, training, education, culture and social and medical assistance;" which directly aligns with SDG 1 (no poverty). The Council of Europe also recognizes the close link between SDGs and the ESC, with a strong emphasis on SDG 1 and SDG 3 (good health and well-being) (Council of Europe, 2018).

4.3 European Pillar of Social Rights

The European Pillar of Social Rights is a framework that includes 20 key principles in order to reach three 2030 headline targets: achieving an employment rate of no less than 78% among those aged 20–64, ensuring that at least 60% of adults take part in training each year, and reducing the number of people exposed to poverty or social exclusion by at least 15 million (European Commission—Directorate-General for Employment, Social Affairs and Inclusion, n.d.). The aims enshrined under the pillar are strongly linked to SDGs related to socio-economic issues.

Financial inclusivity, economic growth, and social equality are the backbone of sustainable development (United Nations, n.d.). Financial inclusion is significant because it is the key to achieving seven out of the 17 SDGs (World Bank, n.d.). Financial inclusion also has paramount importance from the perspective of human rights, such as the right to equality and non-discrimination (Chitimira & Warikandwa, 2023). Albeit financial inclusion is not specifically referred to under the agreement, socio-economic rights are difficult to achieve without financial inclusivity (Chitimira & Mavhuru, 2024).

Despite the role of financial inclusion in achieving socio-economic rights, ECHR, which is the fundamental human rights document for the continent, also does not consist of financial inclusion as a separate human right, and there is no case law of EcrtHR which connects financial inclusion to the fundamental human rights enshrined under the Convention.

As demonstrated by the above-mentioned several international instruments, a strong link between the environmental, social, and economic foundations of SDGs and the obligations under international law exist. Compared with international texts on environmental protection, legally binding documents on socio-economic rights are more general in their wording, rather than specifying specific aims. However, there is no doubt about the binding nature of these obligations, and their violation constitutes a breach of international law (Crawford, 2002, 2019; International Law Commission [ILC], 2001). For this reason, many of the legal documents mentioned above, even if drafted before the concept of SDGs existed, form the legal basis for achieving these goals.

On the other hand, it should be emphasized that the SDGs are not generally binding under international law, but this does not mean there is no expectation on countries to achieve these goals (Ozdamirova et al., 2024). Indeed, the aforementioned objectives are not merely a non-binding international framework, but essentially a set of goals that are fundamental for the survival of the world and the future generations, and it could be argued that the necessity for all countries to strive to achieve these objectives is not contingent upon their being enshrined in a legally binding text (Guiry, 2024b; United Nations, n.d.).

The fact that the above-mentioned international documents do not contain any reference to the contribution of fintech to the SDGs does not diminish the importance of fintech achieving these goals. Likewise, these texts impose international obligations, and how these obligations are to be fulfilled is a matter to be resolved by the decision-making mechanisms of the states.

Having established the legal foundations of state obligations under environmental and socio-economic dimensions, it is important to examine how financial technologies operationalize these obligations. In this regard, the adequacy of the steps taken by states to fulfil these obligations must be assessed within the framework of international texts. Building upon this, the steps taken in countries towards sustainable development will be examined in the following section, focusing on sustainability and fintech.

5 Fintech and the Implementation of SDGs

As the relation between fintech and SDGs is set out under the "Fintech and Sustainable Development Goals" chapter in detail, it is also important to focus on certain fintech concepts linked with SDGs, what kind of actions are taken by states regarding the above-mentioned international legal obligations, and SDGs, which could impact financial technologies that contributes these goals.

Before examining the regulatory steps countries have taken within the scope of the SDGs, it is important to demonstrate that what kind of financial technologies contribute to the achievement of SDGs in practice (Offiong et al., 2025). Likewise, following fintech mechanisms is of paramount importance in achieving SDGs, as they illustrate how fintech channels could operationalize sustainability objectives.

Knowledge of key fintech mechanisms is also important in terms of understanding and being able to critique regulations related to the subject Here is the APA 7th edition citation for the master's thesis (Omeragić, 2023). Up to this date several fintech mechanisms which contribute to the achievement of SDGs were developed, not limited with those mentioned below.

5.1 *Green Finance, Green Fintech, Green Bonds, and Insurtech*

Green finance, referring to using resources for environmentally friendly activities, is an important element of sustainable development (Wang et al., 2022). Currently, the multilateral development banks that shoulder the duty of developing countries prioritize green finance in their activities. For example, the European Bank for Reconstruction and Development (EBRD) made green finance one of its objectives ahead of the 2015 Paris Conference (European Bank for Reconstruction & Development, 2015). The World Bank, the most influential multilateral development bank, started 100% alignment between its activities and the goals of the Paris Agreement as of July 1, 2023 (World Bank Group, 2021). In simple terms, this means that the capital is utilized to fund green projects, and green finance is crucial to development activities.

Such objectives in terms of investment by international financing institutions are most welcome, since according to the UN Environment Programme Finance Initiative, approximately \$5–7 trillion a year until 2030 is necessary to achieve SDGs (UNEP FI, n.d.).

Fintech has paramount importance in advancing green finance, as it stimulates innovative ideas, helps to decrease financial costs, broadens risk management capacity, and enhances income from sustainable financial operations (Xu et al., 2025). However, within a perspective of sustainable development, green fintech seems to be overlooked (Puschmann & Khmarskyi, 2024). The term refers to any kind of financial technology that contributes to overcoming environmental and sustainability challenges (Geetha & Biju, 2024).

According to a study conducted in 2019, shifting the financial structures of all countries worldwide to at least 50% equity financing would reduce total current per capita carbon emissions by approximately 11.5%. This reduction corresponds to approximately one-quarter of the 40% emission reduction that countries under the Paris Agreement have committed to achieve by 2030 (De Haas et al., 2019). This finding underlines why integrating fintech into climate policy could accelerate achieving Paris Agreement targets. Therefore, supporting green fintech which mitigates the effects of climate change and helps achieve the SDGs, is important and necessary.

Yet, green fintech is still at an early venture funding stage (Qiao et al., 2025). In 2024, climate fintech investments increased 17% compared with 2023, totaling 2.7 billion USD (CommerzVentures, 2025). The increasing trend of green fintech attracting funding is a strong signal of its future potential to contributing to the SDGs.

Green bonds are developed to raise funds for green projects and, according to Bai (2025), can be issued by "governments, financial institutions, corporations, or other organizations". 95% of green bond issuers adopt the International Market Association's Green Bond Principles (GBPs) (International Finance Corporation, 2020). While non-compliance with these principles does not constitute a legal deficiency, it may nevertheless undermine credibility of issuers within the market. Green bonds are a direct way to contribute to achieving the SDGs as they are utilized finance green projects (MacAskill et al., 2021).

Insurtech, the abbreviation for insurance technology, is considered an important element of financial inclusion, as it increases accessibility with its lower cost (Wrede, 2018). A research forecasts substantial growth in the insurtech industry, with its size increasing from around USD 36.05 billion in 2025 to over USD 609.50 billion by 2034 (Zoting, 2025).

6 Regulations in the European Union, Germany, and Türkiye with Relation to Sustainable Fintech

The suitability of regulations affects the innovation and progress of sustainable fintech. Similarly, environmental regulations could contribute to the development of green finance instruments, however could also cause financing challenges (Ni et al., 2023). This situation affected environmental regulation in several states (Ni et al., 2022). Therefore, it is necessary to examine what kind of the regulations there are regarding sustainability, which affects development of sustainable fintech.

6.1 EU Legal Framework for Sustainable Finance

The European Union has a well-developed legal framework for sustainable finance (Zetzsche et al., 2022). Many of the steps taken by the EU towards sustainability are aimed at ensuring both the achievement of the SDGs and the fulfilment of obligations under the Paris Agreement (European Commission, 2021b; Oberthür & Dupont, 2021). The importance of examining EU legislation is also evident from the significant share of EU-based fintech companies in global sustainable fintech investment (European Investment Bank, 2024). Furthermore, many jurisdictions outside the EU are modelling their sustainable finance regulations on the EU legal framework (Knaack & Larionova, 2022).

6.1.1 European Green Deal

Sustainable finance can be described as the use of financial instruments to promote economic development while simultaneously limiting environmental harm, in support of the climate and environmental objectives of the European Green Deal and in conjunction with social and governance considerations (European Commission, n.d.). The European Green Deal is a strategy; accordingly, all policies of the EU should aid the achievement of a sustainable future. The Deal also aims to achieve the goals presented in the Paris Agreement, to which the EU and all member states of it are signatories. The European Green Deal includes several legally binding and non-binding tools, such as the EU Climate Change Act and Sustainable Europe Investment Plan (SEIP) (Möslein, 2025).

Although the EU Climate Act does not make a direct reference relating SDGs to the finance and fintech, it refers to the fact that the aims included in Paris Agreement requires capital allocation and investment patterns should be structured to facilitate a shift to a low-carbon, climate-resilient economy (European Parliament & Council of the European Union, 2021; Sikora, 2020). The SEIP constitutes the financial component of the European Green Deal and accordingly digitalization is fundamental for Green Deal albeit there is no reference in the text to the role of fintech or financial technologies in fulfilling the obligations under Paris Agreement (European Commission, 2020).

6.1.2 EU Taxonomy Regulation

In order to reach the objectives set out under the European Green Deal, the "EU Taxonomy" was created and came into force in July 2020 (Möslein, 2025). The taxonomy regulation sets out a classification system for sustainable economic activities (Fuest & Meier, 2023). According to this regulation, an economic activity must meet four conditions to be considered environmentally sustainable. These are, sustainable use and protection of water and marine resources, the transition to a circular economy, pollution prevention and control, and the protection and restoration of biodiversity and ecosystems (van den Hurk & van der Klooster, 2024).

According to the European Commission (n.d.), determining which activities are 'sustainable' is essential to achieving the EU's environmental and energy targets, and within this scope, a common classification system called the 'EU Taxonomy' has been developed. It is envisaged that the system will prevent dangers such as 'greenwashing', create unity in the market regarding sustainability, and ensure that companies carry out environmentally conscious activities (Och, 2025).

Under the EU Taxonomy Regulation, unlike other regulations, there is a specific reference to the importance of developing environmentally sustainable financial instruments in directing private capital towards sustainability (European Parliament & Council of the European Union, 2020; Papari et al., 2024; Schütze & Stede, 2021). This reference confirms the role of financial technologies in achieving SDGs.

6.1.3 Sustainable Finance Disclosure Regulation (SFDR)

Directly linked with fintech, the Sustainable Finance Disclosure Regulation (SFDR), which entered into force in March 2021, requires transparency on sustainability risk management from financial market actors (European Commission, 2023; Steuer & Tröger, 2022).

Although SFDR is helpful, it requires SMEs and small companies to manage costly disclosure processes by themselves (Rossmann et al., 2025). Additionally, it poses challenges for non-EU companies seeking to enter the market (Sahintürk, 2025). Companies that lack the resources to comply with disclosure obligations would naturally be excluded from the market (Ortiz-Martínez & Marín-Hernández, 2024), thereby negatively affecting the development of fintech, which could contribute to the SDGs.

Under SFDR, there is no requirement to eliminate the adverse effects of climate change. Other negative aspects include a lack of clarity on key elements of the SFDR and and self-assessment by companies, which is prone to greenwashing (Busch, 2023). Certain disclosure obligations are also irrelevant for different industries, creating unnecessary bureaucracy for those companies (European Parliament, 2024). Although SFDR obliges companies to disclose information on their sustainable activities, it does not monitor whether the claims they make are factual (European Commission, 2023).

6.1.4 Corporate Sustainability Due Diligence Directive (CS3D)

Management of supply chains has utmost importance for decarbonization, lowering carbon emissions (Mubarik, 2025), and consecutively achieving the goals set out under the Paris Agreement. EU's Corporate Sustainability Due Diligence Directive (CS3D), which is applicable extraterritorially, entered into force on July 25, 2024, and not only requires disclosure of adverse sustainability effects but also to prevent it (Sinnig & Zetzsche, 2025). EU companies with more than EUR 450 million in global net turnover and over 1,000 employees, and non-EU companies with more than EUR 450 million in turnover in the EU, are within the scope (Baker McKenzie, 2024; European Parliament & Council, 2024). However, the due diligence duty required under the Directive was not new for certain member states, namely France and Germany.

In France, the Duty of Vigilance Law entered into force on March 27, 2017. Similar to CS3D, it requires French companies with more than 5.000 employees (or 10.000 for companies with subsidiaries abroad) to prepare and implement a plan to mitigate human rights violations, including environmental damage (Cossart et al., 2017; France, 2017; Schilling-Vacaflor, 2021). Although regarding the employment rights, on 5 December 2023, France's privatized national postal service *La Poste*, stands as the first company that got a judgment against them for violating the Vigilance Law (Bounfour & Chatelain, 2025). This represents how the obligations of states under international law are reflected and given binding effect in domestic legislation.

Similarly, Germany's Supply Chain Act which entered into force on January 1, 2023 regulates the responsibility of German enterprises with respect to their global supply chain (Mittwoch & Bremenkamp, 2023). Cases on alleged forced labor in their supply chains in the Xinjiang Uyghur Autonomous Region were filed against German manufacturers Volkswagen, BMW, and Mercedes-Benz on June 21, 2023. According to the applicant, the European Center for Constitutional and Human Rights (ECCHR), the Federal Office for Economic Affairs and Export Control (BAFA) did not share information about ongoing proceedings regarding the complaint due to security risks and the potential oppression of individual workers (European Center for Constitutional and Human Rights, 2024).

In case of a breach, France's Vigilance Law and Germany's Supply Chain Act impose respectively financial penalties and administrative fines (Brabant & Savourey, 2017; Mittwoch & Bremenkamp, 2023). That is a fundamental difference with CS3D, where companies are required to bear full civil liability and the necessity to compensate victims (Sinnig & Zetzsche, 2025).

Although it entered into force, in order for CS3D to be applicable, there two years of transition period, where member states are required to integrate the directive to their national laws (Feigerlová, 2024). However, the integration of the directive into national legislation is currently seen as a challenging journey (Mieszkowska, 2024). Indeed, two heads of state, Chancellor Merz for Germany and President Emmanuel Macron, called for CS3D to be taken off the table in May 2025 (ESG Today, 2025; Thorens et al., 2025). Contests against the law are not limited to the governments. The CEOs of 46 companies, including Siemens and TotalEnergies, sent a letter to the EU requesting that the CS3D be scrapped altogether, citing uncertainty in the sustainability reporting process (FIDH & FIR, 2025; Williment, 2025).

6.1.5 Corporate Sustainability Reporting Directive (CSRD)

The Corporate Sustainability Reporting Directive (CSRD) was created to as a result of the need to update the Non-Financial Reporting Directive (NFRD) (Martinčević et al., 2024; Radu et al., 2023). The directive imposed an annual reporting obligation on companies to report annually on their sustainability activities. However, after it became apparent that the information provided by companies did not contain the information deemed important by investors and other stakeholders (Radu et al., 2023), the decision was made to update the directive (Dunfjäll, 2025; European Commission, 2022). One of the most fundamental differences between the NFRD and the CSRD is that the latter covers all large companies, regardless of whether they exceed the 500-employee threshold (Martinčević et al., 2024). The CSRD also covers SMEs, except for micro-enterprises. Thus, the aim is to ensure that investors to have all the necessary information about companies, particularly regarding sustainability (European Commission, 2021a; Monciardini, 2025).

6.1.6 EU Green Bond Standard Regulation (EU GBS)

The EU Green Bond Standard Regulation (EU GBS) establishes common criteria for companies issuing green bonds in the EU (Pyka, 2023). The regulation, which came into force on December 20, 2023, includes numerous reporting and reviewing obligations in accordance with the principle of transparency (Brückbauer et al., 2023). These include the obligation for issuers to obtain approval from an external reviewer before issuing green bonds and to have the bonds reviewed externally after issuance (Lukács, 2025).

Given the foregoing, compared to SFDR, EU GBS is more resilient to greenwashing, as it includes an external review mechanism (Koch, 2023; Lehmann, 2021); however, it still does not address the problem of the non-inclusion of SMEs and fintech companies that contribute to sustainable development (Badenhoop, 2022). These companies often work with limited resources, yet are obliged to fulfil several additional obligations to issue green bonds (Deschryver & de Mariz, 2020). Although EU GBS includes a review by December 21, 2028, to consider SMEs' market share (Özçelik et al., 2025), it does not include any projections on how this situation will be addressed if SMEs' market share remains very limited.

6.1.7 The Omnibus Simplification Package and Regulatory Simplification

Given the complexity of the regulations outlined above and the variety of obligations they impose on companies, the EU published a simplification package, known as the Omnibus Simplification Package, on February 26, 2025. This package covers many areas, including sustainable finance, and aims to attract investment and increase competition by simplifying regulations. The package sets a target of substantially easing regulatory obligations, with an overall 25% reduction in administrative burdens for all businesses and a higher 35% relief for SMEs (European Commission, Directorate-General for Financial Stability, Financial Services and Capital Markets Union, 2025).

The package also envisages amendments to the CSRD and CS3D regulations. The proposed amendments to the CSRD with the Omnibus Package aimed to exclude 80% of companies from its scope. However, the package proposal was rejected by the European Parliament on 22 October 2025 and consequently negotiated from scratch and proposed again (Genovese, 2025). After a year of uncertainty, it was finally greenlit by the European Parliament with the change of decreasing the scope of CSRD by %90 (Costa, 2025).

The Omnibus Simplification Package is significant for understanding the EU's regulatory process regarding sustainability. Indeed, recalling the fact that the number of companies falling within the scope of the NFRD has been increased as a result of the CSRD, and now that number is decreased with the Omnibus Simplification Package, it can be argued that the envisaged changes reveal that normative stability has not yet been achieved in the EU's sustainability policies and that the regulatory process has a reactive and experimental character rather than a holistic vision (Bertram, 2025; de Sadeleer, 2025). Additionally, the lack of data concerning SMEs and consequently a substantial number of fintech companies as a result of the amendment package may negatively affect improvement of the regulatory framework (Financial Stability Institute, 2024).

Taken together, the EU's sustainable finance rules shape operating environment of fintech in the region (Giakoumelou et al., 2024). These regulations were made by the EU in the context of international obligations relating to the SDGs, and they are important for understanding how international texts containing general obligations, such as the SDGs, are reflected in domestic law.

6.2 Germany's National Approach to Fintech Regulation

Above-mentioned legislation regarding sustainable finance does not include clauses which support sustainable fintech directly. Similarly, Germany's draft law currently under discussion, the Second Financing for the Future Act (Entwurf eines Zweiten Gesetzes zur Finanzierung von zukunftssichernden Investitionen (Zweites Zukunfts-finanzierungsgesetz – ZuFinG II)) also does not include specific reference to fintech, and its important role in achieving SDGs (Bundesrat, 2024).

Despite the lack of recognition of role of fintech on achieving the SDGs in its legislation, Germany, the second-largest financial center in Europe and a leading fintech hub on the continent (Hornuf & Mattusch, 2025), has taken additional steps to support the fintech sector. For example, the "FinTech Made in Germany" initiative made the Federal Financial Supervisory Authority's (BaFin) assistance available to fintech companies to support their disclosure obligations (Federal Ministry of Finance, 2024).

6.3 Türkiye's National Framework on Sustainable Finance

In Türkiye, the Environment Law No. 2872 states that all parties are responsible for the protection of the environment, however, as this act primarily covers waste management, it is not directly linked to either fintech or sustainable development (Environment Law No. 2872 (Türk.), 1983).

Despite the lack of reference to the role of fintech, the recently adopted Climate Law contains provisions stipulating green and sustainable capital market instruments, and encourages the activities and investments to be undertaken by institutions and organizations to combat climate change (Climate Law No. 7552 (Türk.), 2025). Under the law, in order to contribute to climate financing, establishment of the Türkiye Green Taxonomy by the Directorate of Climate Change is also envisaged (Climate Law No. 7552 (Türk.), 2025).

General Provisions on the Disclosure of Financial Information Related to Sustain-ability in Türkiye takes EU's CSRD as basis, and impose reporting obligations on companies regarding their sustainable activities (Kamu Gözetimi, Muhasebe ve Denetim Standartları Kurumu, 2023).

Türkiye's exchange entity Borsa Istanbul's (BIST) sustainability index commenced in 2014, where sustainability assessment of volunteering companies is being made in cooperation with London Stock Exchange Group (LSEG) (Borsa İstanbul, n.d.). In order to be included in the BIST Sustainability Index, a company voluntarily needs to demonstrate that it meets three criteria, including passing a threshold in activities across the three pillars of the SDGs (Tanç & Çardak, 2024). As the banks lead the sectoral distribution of the BIST Sustainability 25 once again highlights the importance of sustainable finance (Borsa İstanbul A.Ş., n.d.).

However, the integration of financial technologies with the SDGs has not yet been fully achieved (Bayram et al., 2022). Similarly, the 'Türkiye Fintech Ecosystem Status Report' published by the Presidential Finance Office for 2024 merely states that companies developing products and services that meet environmental, social, and corporate governance criteria are leading the field of financial technologies (T.C. Cumhurbaşkanlığı Finans Ofisi, 2025). Indeed, the report emphasizes that digitalization is particularly focused on the banking sector, rather than the SDGs itself.

On the other hand, Türkiye's Development Plan for 2024–2028 states that financing will be provided for green technological investments aimed at reducing the carbon footprint to zero and that Türkiye is expected to become one of the leading countries in global green finance. The plan also states that, under the leadership of the Istanbul Financial Centre (IFC), the development of financial technologies will be ensured, and Türkiye will be among the leading countries in terms of the effectiveness of fintech activities. The ambitious plan also includes elements such as supporting fintech companies that operate on a participatory basis, to establish an inclusive financial ecosystem, thereby contributing to SDGs, despite not directly mentioning the connection (T.C. Cumhurbaşkanlığı Strateji ve Bütçe Başkanlığı, 2023).

6.4 Financial Inclusion in Comparative Perspective

According to the World Bank's Global Findex Database, while in the Euro area the percentage of people above age fifteen who have a bank account or used a mobile money service in the past year is around 98% for both genders, for low-income countries the percentage descends to 43.9% for males and as low as 34.5% for females (Demirguc-Kunt et al., 2022). Both gender and number discrepancies are astonishing, and it could thus be said that creating regulations which consider inclusivity is more important in developing countries than in developed ones (Chen & Divanbeigi, 2019; Sharma & Kukreja, 2013). However, there is no specific regulation directly mentions financial inclusion in Türkiye, an emerging economy, and the situation is similar in the EU.

The absence of references to financial inclusion in domestic legislation is also considered a deliberate move, as the balance between financial inclusion and financial stability reflects policymakers' reservations about expanding social financial access while minimizing systemic risks (Martins Pereira & Lima, 2024). Despite this, there are policy interventions by developed countries like Germany in order to advance financial inclusion (Neuberger, 2015; Yoshino & Morgan, 2016). This is mostly realized through the regulations regarding microfinance institutions (Neuberger, 2015).

The situation is similar under international law, where financial inclusion is not explicitly referred to under legally binding international documents (G20 Financial Inclusion Experts Group, 2015). This is expected, as international law is driven by developed countries (Ikejiaku, 2014), where financial inclusion is not a prominent problem compared with the emerging economies and developing countries (Demirguc-Kunt et al., 2022). The lack of legal documents under international law related to financial inclusion once again underscores the importance of including several types of economies to the creation of international law.

This demonstrates that despite progress in digitalization, the legal framework has yet to institutionalize inclusion as a right or state obligation (Lima & Pereira, 2024). Therefore, in both the international and domestic regulatory contexts, the environment has taken precedence over the social and economic pillars of the SDGs (Khot et al., 2026). This is also reflected in fintech mechanisms: those related to the environment, such as green finance, green fintech, and green bonds, are numerically more than the financial inclusion tools, such as insurtech.

7 Future of Sustainable Fintech Governance

Regardless of the hurdles that certain regulations bring, the need for governments to have information to promote the industry and to make targeted regulations is undeniable (Barefoot, 2020; Naseem & Alim, 2026). Effectiveness of regulatory framework is a fundamental component of success of fintechs (Harsono & Suprapti, 2024). However, as mentioned above this should not be at the cost of companies that are unable to navigate through the legal requirements.

Against this background, a hybrid governance system that combines formal and informal institutions in a framework might be a solution. Hybrid governance aims to strengthen the official framework for sustainability by creating synergy between the state regulators and stakeholders (Colona & Jaffe, 2016; Wilson et al., 2025) This way, stakeholders, who are the recipients of the regulatory framework will be more involved in the process, rather than only be subjected certain obligations arises from the regulations. States and respective companies should improve governance systems in order to support sustainable development (Liu & Zhang, 2024) and hybrid governance could be a way to achieve this. It is important to emphasize that hybrid systems should not create new hurdles for fintech companies, especially SMEs, which contribute to achieving the SDGs.

Economic incentives could be another idea to consider while constructing sustainability governance. Investing in the green finance sector is considered more costly (Gavril, 2015) and has a longer payback period compared to other sectors (Inderst et al., 2012). Especially for emerging markets, incentives have paramount importance to support green finance (Okeke et al., 2024).

Additionally, it is believed that financial incentives will play a significant role in ensuring a sustainable future (Fu et al., 2023). Drawing a comparison between green finance and sustainable fintech, it can be said that incentives will play an important role in the development of sustainable fintech as well. Therefore, similar to Germany, incentives to those disadvantaged companies in the field of sustainable fintech could be another way to move forward with the sustainable fintech governance (Federal Ministry of Finance, 2024).

Another perspective that requires attention is the need to create global standards to ensure fintech solutions that contribute to sustainable development are credible and easy to operate cross-border. Green finance technologies, especially those aimed at individual consumers, often offer services that require cross-border transactions (He, 2021). However, regulatory differences between countries regarding sustainability constitute an important hurdle (OECD, 2023). The other cross-border operation challenges include data privacy, legal issues such as jurisdiction, and cybersecurity (Sharma et al., 2021).

Cross-border operational obstacles may result in duplicated compliance obligations, increased difficulty arising from unfamiliar legal frameworks, and potential reputational harm due to differing regional sustainability classifications. Additionally, cross-border operation obstacle may cause companies operate in countries where sustainability rules are more lenient, leading to greenwashing (Gauer, 2025).

Establishing international standards for regulations to ensure the smooth operation of cross-border transactions will prevent bottlenecks in regulations and contribute to the expansion of financial technologies (Bhattacharya & Sachdev, 2024).

Against this background, the following conclusion draws together the key findings of this chapter and outlines their implications for the future of sustainable finance governance.

8 Conclusion

The SDGs are essentially an international framework containing a number of goals based on environmental, social and economic principles. Although this framework is not legally binding under international law, the goals set out therein are included in legally binding instruments of international law. Thus, a large part of the SDGs constitutes an obligation that countries must fulfil.

In this connection, the Paris Agreement stands out in terms of environmental SDGs, while general international legal texts adopted even before the SDGs were established stand out in terms of socio-economic SDGs.

International obligations of the states are manifested as domestic legal regulations. Against this background, the extent to which obligations regarding SDGs are fulfilled, could be understood by examining domestic regulations.

Considering the environmental, social and economic contributions of fintech, it can be said that their role in fulfilling the above-mentioned obligations is undeniable. The fintech companies in question are subject to sustainability regulations in their respective countries and must operate within this ecosystem.

The EU plays a leading role in the domestic legal manifestation of the SDGs in terms of regulation. The EU's regional regulations have a direct impact on member states. Germany, an EU member state and one of the region's financial centers, is striving to create an environment conducive to the development of sustainable fintech through its own domestic regulations, in addition to those of the EU. However, lack of information regarding the breach processes of sustainability laws poses a threat to information acquisition and consequently effectiveness of the legislation. Türkiye, as an emerging market, follows leading regulators such as the EU, however, envisages steps to be taken within its own borders to achieve SDGs through fintech.

In general, it is observed that legislators take reactions into account while drafting regulations affecting many sectors including fintech, particularly in terms of sustainability. In regulations related to sustainability, which reflect international obligations under the SDGs in domestic law, it is understood that the role of fintech is not specifically addressed and that there is no consideration of the tailor-made framework for the development of sustainable fintech.

Although there is currently no clear trend in sustainability towards supporting specific sectors such as fintech through regulations, it is evident that more steps need to be taken to fulfil international obligations in this regard. This is clear, as the Paris Agreement's target of limiting global warming to 1.5 °C above pre-industrial levels was not achieved for 2024, and 2024 has gone down in history as the first year in which global warming exceeded such a threshold (World Meteorological Organization, 2025).

Despite exceeding the global warming limit recorded in the Paris Agreement in 2024, environmental issues are at the forefront of the EU's sustainability regulations. Environmental protection is also a key SDG in the mechanisms developed in the fintech sector. One of the main reasons for this situation is the existence of a legally binding text, such as the Paris Agreement, which contains material and specific targets related to environmental protection, and attributes the achievement of these targets as an international legal obligation.

The other two pillars of the SDGs, namely social and economic rights do not share the same fate. Financial inclusion is a concept that is closely tied to socio-economic SDGs. However, its absence of mention as a state obligation under international law reflects in the same way to domestic legal regulations. The vast difference between environmental protection and financial inclusion in terms of regulations and fintech innovations, clearly demonstrates the impact of international obligations.

Another obstacle to fulfilling international obligations through fintech in the context of SDGs is the lack of cross-border cooperation in regulations. Indeed, recent fintech startups focusing on investments, such as Robinhood, are regional—a hurdle that could be overcome by adopting a global regulatory framework or cross-border cooperation. This would ensure people around the world—especially in developing countries—have access to global investment opportunities, eventually contributing to financial inclusivity, economic growth, and social equality.

In conclusion, it is clear that sustainable finance plays a fundamental role in achieving the SDGs. Against this background, fintech is an important tool for states to fulfil their international obligations related to the SDGs. To achieve this, the establishment of a regulatory framework to develop and ensure the continuity of fintech is necessary. Indeed, it is expected of states to use their governance to take actions in relation to their international and national legal obligations. In the fintech sector, this obligation can be fulfilled by assessing the shortcomings of existing regulations and other possible steps in a dynamic and innovative manner.

References

Arnardóttir, O. M. (2017). Res Interpretata, Erga Omnes effect and the role of the margin of appreciation in giving domestic effect to the judgments of the European Court of Human Rights. *European Journal of International Law, 28*(3), 819–861.

Badenhoop, N. (2022). *Enhancing the European green bond standard: Regulatory options and challenges for market uptake*. European Parliament Policy Department for Economic, Scientific and Quality of Life Policies.

Bai, Y. (2025). The impact of green bond issuance on corporate environmental and financial performance: An empirical study of Japanese listed firms. *International Journal of Financial Studies, 13*(3), 141. https://doi.org/10.3390/ijfs13030141

Baker McKenzie. (2024). *European Union: The new corporate sustainability due diligence directive—What does it mean for companies?* Baker McKenzie.

Barefoot, J. (2020). *Digitizing financial regulation: Regtech as a solution for regulatory inefficiency and ineffectiveness* (M-RCBG Associate Working Paper No. 150). Mossavar-Rahmani Center for Business & Government, Harvard Kennedy School.

Bayram, O., Talay, I., & Feridun, M. (2022). Can fintech promote sustainable finance? Policy lessons from the case of Turkey. *Sustainability, 14*(19), 12414. https://www.mdpi.com/2071-1050/14/19/12414

Bennouna, M. (2017). Le droit international entre la lettre et l'esprit: Cours général de droit international public. *Recueil des Cours De L'académie De Droit International De La Haye, 383*, 9–91.

Bernard, F., & Da Rugna, A. (2024). *La qualité de victime en matière climatique selon la CourEDH*. Sui-Generis. https://doi.org/10.21257/sg.268

Bertram, A. (2025). Deregulating to no avail: How the Omnibus package falls short in simplifying key EU green deal instruments. *Intereconomics: Review of European Economic Policy, 60*(3), 170–175. https://doi.org/10.2478/ie-2025-0033

Bhattacharya, S., & Sachdev, B. K. (2024). Regulatory challenges and opportunities for future fintech and green finance. In V. Sharma, M. Gupta, N. Arora, & A. Rocha (Eds.), *Blockchain's transformative potential of financial technology for sustainable futures*. Information Systems Engineering and Management (Vol. 17). Springer. https://doi.org/10.1007/978-3-031-70219-8_7

Borsa İstanbul, A. Ş. (n.d.). *BIST Sustainability 25 (XSD25) index*. Retrieved October 31, 2025, from https://www.borsaistanbul.com/en/index/xsd25

Borsa İstanbul, A. Ş. (n.d.). *Sustainability—BIST stock indices*. https://www.borsaistanbul.com/en/indices/bist-stock-indices/sustainability

Bounfour, T., & Chatelain, L. (2025). A (so far unlikely) day in court: An overview of the first judicial decisions under the French Duty of Vigilance Law. *Business and Human Rights Journal, 10*, 434–439. https://doi.org/10.1017/bhj.2025.22

Brabant, S., & Savourey, E. (2017). *France's corporate duty of vigilance law: A closer look at the penalties faced by companies*. Revue Interna-tionale De La Compliance Et De L'éthique Des Affaires 50.

Brückbauer, F., Cézanne, T., Kirschenmann, K., & Schröder, M. (2023). *Does the European Union need another green bond standard?* (ZEW Policy Brief No. 10/2023). ZEW – Leib-niz-Zentrum für Europäische Wirtschaftsforschung.

Bundesrat. (2024, December 4). *Entwurf eines Zweiten Gesetzes zur Finanzierung von zukunftssichernden Investitionen (Zweites Zukunfts-finanzierungsgesetz – ZuFinG II)* [Draft law, Drucksache 599/24]. https://www.bundesrat.de/SharedDocs/drucksachen/2024/0501-0600/599-24.pdf?__blob=publicationFile&v=1

Busch, D. (2023). EU sustainable finance disclosure regulation. *Capital Markets Law Journal, 18*(3), 303–328. https://doi.org/10.1093/cmlj/kmad005

Chatjuthamard, P., Chintrakarn, P., Jiraporn, P., & Lee, S. M. (2025). Climate change, tax avoidance, and shareholder value: Evidence from the Paris Agreement. *International Review of Finance, 25*(1), 1–10. https://doi.org/10.1111/irfi.12454

Chen, R., & Divanbeigi, R. (2019). *Can regulation promote financial inclusion?* (World Bank Policy Research Working Paper No. 8711). World Bank.

Chitimira, H., & Mavhuru, L. (2024). Towards the recognition of financial inclusion as a fundamental socio-economic right in selected SADC countries. *South African Mercantile Law Journal, 36*(3), 440–456. https://doi.org/10.47348/SAMLJ/v36/i3a5

Chitimira, H., & Warikandwa, T. V. (Eds.). (2023). *Financial inclusion and digital transformation: Regulatory practices in selected SADC countries: South Africa, Namibia, Botswana and Zimbabwe* (Ius Gentium: Comparative Perspectives on Law and Justice, Vol. 106). Springer Cham. https://doi.org/10.1007/978-3-031-23863-5

Climate Law No. 7552, Official Gazette No. 32951 (Türkiye, 2025).

Colona, F., & Jaffe, R. (2016). Hybrid governance arrangements. *European Journal of Development Research, 28*, 175–183. https://doi.org/10.1057/ejdr.2016.5

CommerzVentures. (2025). *Climate fintech report 2025*. https://www.commerzventures.com/

Cossart, S., Chaplier, J., & de Lomenie, T. B. (2017). The French law on duty of care: A historic step towards making globalization work for all. *Business and Human Rights Journal, 2*(2), 317–323.

Costa, M. (2025, December 18). *EU sustainability omnibus package approved, ESRS data points reduced*. Green Central Banking. https://greencentralbanking.com/2025/12/18/eu-sustainability-omnibus-package-approved-esrs-data-points-reduced/

Council of Europe. (2018, March 23). *Council of Europe contribution to the UN 2030 Agenda for sustainable development* [Rolling document]. https://rm.coe.int/coe-contribution-to-the-agenda-2030-rolling-document/16807959a6

Council of Europe. (2025). *Signatures & ratifications—European social charter* [Web page]. https://www.coe.int/en/web/european-social-charter/signatures-ratifications

Crawford, J. (2002). *The international law commission's articles on state responsibility: Introduction, text and commentaries*. Cambridge University Press.

Crawford, J. (2019). *Brownlie's principles of public international law* (9th ed.). Oxford University Press.

Davis, D. M. (2024). Socio-economic rights. In M. Rosenfeld & A. Sajo (Eds.), *The Oxford handbook of comparative constitutional law* (chap. 49). Oxford University Press. https://doi.org/10.1093/law/9780199578610.001.0001/law-9780199578610-chapter-51

Degan, V. D. (1997). *Sources of international law* (Vol. 00027). Brill | Nijhoff.

De Haas, R., Martin, R., Muûls, M., & Schweiger, H. (2019). *Finance and carbon emissions*. European Bank for Reconstruction and Development. https://www.ebrd.com/publications/working-papers/finance-and-carbon-emissions

de Sadeleer, N. (2025). The European green deal: Greenwashing compounded by deregulation (Omnibus law) or a genuine paradigm shift? *European Journal of Risk Regulation, 16*(4), 1343–1374. https://doi.org/10.1017/err.2025.20

Demaria, T. (2021). Obligations de comportement et obligations de résultat dans la jurisprudence de la Cour internationale de Justice. *The Canadian Yearbook of International Law/Annuaire canadien de droit international, 58*, 362–381. https://hal.science/hal-03524958

Demirguc-Kunt, A., Klapper, L., Singer, D., Ansar, S., & Hess, J. (2022). The global Findex database 2021: Financial inclusion, digital payments, and resilience in the age of COVID-19. *World Bank*. https://doi.org/10.1596/978-1-4648-1897-4

Deschryver, P., & de Mariz, F. (2020). What future for the green bond market? How can barriers to scalability be removed? *Journal of Risk and Financial Management, 13*(3), 61. https://doi.org/10.3390/jrfm13030061

Dodson, J., Derer, P., Cafaro, P., & Gotmark, F. (2022). Population growth, family planning and the Paris Agreement: An assessment of the Nationally Determined Contributions (NDCs). *International Environmental Agreements: Politics, Law and Economics, 22*(3), 561–576. https://doi.org/10.1007/s10784-022-09573-8

Dunfjäll, M. (2025). Materiality in transition: Challenges and opportunities in corporate sustainability reporting under the CSRD. *European Journal of Risk Regulation, 16*(3), 927–941. https://doi.org/10.1017/err.2025.10016

Environment Law No. 2872, Official Gazette No. 18132 (Türkiye, 1983).

ESG Today. (2025, May 20). *France's Macron joins Germany's call to scrap EU's supply chain sustainability due diligence law.* https://www.esgtoday.com/frances-macron-joins-germanys-call-to-scrap-eus-supply-chain-sustainability-due-diligence-law/

European Bank for Reconstruction and Development. (2015). *EBRD and green finance.* https://www.ebrd.com/news/2015.html

European Center for Constitutional and Human Rights (ECCHR). (2024, May 27). *German economic engine roars thanks to forced labour: Complaint filed against VW, BMW and Mercedes-Benz* [Web page]. https://www.ecchr.eu/en/case/german-economic-engine-roars-thanks-to-forced-labor-complaint-filed-against-vw-bmw-and-mercedes-benz/

European Commission. (2020, January 14). *Communication from the Commission to the European Parliament, the Council, the European Economic and Social Committee and the Committee of the Regions: Sustainable Europe Investment Plan—European Green Deal Investment Plan* (COM (2020) 21 final). https://eur-lex.europa.eu/legal-content/EN/TXT/?uri=CELEX%3A52020DC0021

European Commission. (2021a, April 21). *Questions and answers: Corporate Sustainability Reporting Directive proposal [Q&A].* https://ec.europa.eu/commission/presscorner/detail/en/qanda_21_1806/

European Commission. (2021b). *Strategy for financing the transition to a sustainable economy.* https://finance.ec.europa.eu/publications/strategy-financing-transition-sustainable-economy_en

European Commission. (2022, January 21). Proposal for a Directive of the European Parliament and of the Council amending Directive 2013/34/EU as regards corporate sustainability reporting [COM/2021/0390 final].

European Commission. (2023). *Sustainable finance disclosure regulation (SFDR).* https://finance.ec.europa.eu/sustainable-finance/disclosures/sfdr_en

European Commission. (n.d.). *EU taxonomy for sustainable activities.* Finance—European Commission. https://finance.ec.europa.eu/sustainable-finance/tools-and-standards/eu-taxonomy-sustainable-activities_en

European Commission. (n.d.). *Overview of sustainable finance.* Finance—European Commission. https://finance.ec.europa.eu/sustainable-finance/overview-sustainable-finance_en

European Commission—Directorate-General for Employment, Social Affairs and Inclusion. (n.d.). *European pillar of social rights—Building a fairer and more inclusive European Union* [Web page]. Retrieved October 30, 2025, from https://employment-social-affairs.ec.europa.eu/policies-and-activities/european-pillar-social-rights-building-fairer-and-more-inclusive-european-union_en

European Commission, Directorate-General for Financial Stability, Financial Services and Capital Markets Union. (2025, April 1). *Omnibus package* [News article]. Finance—European Commission. https://finance.ec.europa.eu/news/omnibus-package-2025-04-01_en

European Court of Human Rights. (2024, April 9). *Verein KlimaSeniorinnen Schweiz and Others v. Switzerland* (Application No. 53600/20) [Grand Chamber judgment]. HUDOC. https://hudoc.echr.coe.int/eng/?i=002-14304

European Investment Bank (EIB). (2024). *European Fintech landscape report 2024*. https://www.eib.org

European Parliament & Council of the European Union. (2020, June 18). Regulation (EU) 2020/852 of the European Parliament and of the Council of 18 June 2020 on the establishment of a framework to facilitate sustainable investment and amending Regulation (EU) 2019/2088 [Regulation]. *Official Journal of the European Union Law, 198*, 13–43. https://eur-lex.europa.eu/legal-content/EN/TXT/?uri=CELEX:32020R0852

European Parliament & Council of the European Union. (2021, June 30). *Regulation (EU) 2021/1119 of the European Parliament and of the Council of 30 June 2021 establishing the framework for achieving climate neutrality and amending Regulations (EC) No 401/2009 and (EU) 2018/1999 ("European Climate Law") [Regulation]. *Official Journal of the European Union Law, 243*, 1–17. https://eur-lex.europa.eu/legal-content/EN/TXT/?uri=CELEX:32021R1119

European Parliament & Council. (2024). Directive (EU) 2024/1760 of the European Parliament and of the Council of 13 June 2024 on corporate sustainability due diligence and amending Directive (EU) 2019/1937 and Regulation (EU) 2023/2859 (OJ L 1760, 5.7.2024

European Parliament resolution of 17 December 2020 on sustainable corporate governance (2020/2137(INI)) [2021] C 445/94.

European Parliament. (2024). *The shortcomings of SFDR in practice*. https://www.europarl.europa.eu/RegData/etudes/STUD/2024/754212/IPOL_STU(2024)75412_EN.pdf

European Union. (2023, October 16). *Update of the NDC of the European Union and its Member States [PDF]*. https://unfccc.int/sites/default/files/NDC/2023-10/ES-2023-10-17%20EU%20submission%20NDC%20update.pdf

Federal Ministry of Finance. (2024). *FinTech made in Germany initiative*. https://www.bundesfinanzministerium.de/Content/EN/Standardartikel/Press_Room/Publications/Brochures/fintech-made-in-germany.html

Feigerlová, M. (2024). Moving from pledges to commitments: Analysing climate transition plans in the EU proposal for a corporate sustainability due diligence directive. *Carbon and Climate Law Review, 18*(1), 17–30.

Financial Stability Institute. (2024). *Financial stability implications from FinTech* (BIS Report). Bank for International Settlements.

France. Loi n° 2017–399 du 27 mars 2017 relative au devoir de vigi-lance des sociétés mères et des entreprises donneuses d'ordre, J.O. n° 0074 (28 mars 2017) (Fr.). https://www.legifrance.gouv.fr/jorf/id/JORFTEXT000034290626/

Fu, C., Lu, L., & Pirabi, M. (2023). Advancing green finance: A review of sustainable development. *Digital Economy and Sustainable Development, 1*, Article 20. https://doi.org/10.1007/s44265-023-00020-3

Fuest, C., & Meier, V. (2023). Green finance and the EU-taxonomy for sustainable activities: Why using more direct environmental policy tools is preferable. *The Economists' Voice, 19*(2), 261–266. https://doi.org/10.1515/ev-2022-0021

G20 Financial Inclusion Experts Group. (2015). *Global standard-setting bodies and financial inclusion: Consultation document* (pp. xiii). G20.

Gauer, K. (2025). Comparative analysis of the EU and South African green taxonomies and their legal and financial implications for cross-border investments. In C. Hugo (Ed.), *Annual banking law update 2025—Recent legal developments of special interest to banks*. Juta and Company.

Gavril, I. (2015). *Incentives for financing sustainable development*. Annals of the „Constantin Brâncuşi". University of Târgu Jiu, Economy Series.

Geetha, S., & Biju, A. V. N. (2024). Is green FinTech reshaping the finance sphere? Unravelling through a systematic literature review. *Environmental Science and Pollution Research, 31*, 1790–1810. https://doi.org/10.1007/s11356-023-31382-y

Genovese, V. (2025, October 22). *European Parliament rebels against simplified sustainability requirements*. Euronews—My Europe. https://www.euronews.com/my-europe/2025/10/22/european-parliament-rebels-against-simplified-sustainability-requirements/

Giakoumelou, A., Salvi, A., Bekiros, S., & Onorato, G. (2024). ESG and FinTech funding in the EU. *Research in International Business and Finance, 69*, Article 102233. https://doi.org/10.1016/j.ribaf.2024.102233

Grynchak, A. A., Tavolzhanska, Y. S., Grynchak, S. V., Smorodynskyi, V. S., & Latysh, K. V. (2022). *Convention for the protection of human rights and fundamental freedoms as a constitutional instrument of European Public Order*. Public Organization Review.

Guiry, N. (2024a). *International law and the sustainable development goals*. Snapshots of Doctoral Research at University College Cork.

Guiry, N. (2024b). Why do states adhere to the sustainable development goals? *Transnational Environmental Law, 13*(2), 400–427. https://doi.org/10.1017/S2047102524000190

Hák, T., Janoušková, S., & Moldan, B. (2016). Sustainable development goals: A need for relevant indicators. *Ecological Indicators, 60*, 565–573. https://doi.org/10.1016/j.ecolind.2015.08.003

Harris, D. (2021). *The European social charter: The protection of economic and social rights in Europe* (Vol. 00025, 2nd ed.). Brill|Nijhoff.

Harsono, I., & Suprapti, I. A. P. (2024). The role of fintech in transforming traditional financial services. *Accounting Studies and Tax Journal (COUNT), 1*(1), 81–91.

He, D. (2021). Digitalization of cross-border payments. *China Economic Journal, 14*(1), 26–38. https://doi.org/10.1080/17538963.2020.1870272

Hornuf, L., & Mattusch, M. (2025). Fintech startups in Germany: firm failure, funding success, and innovation capacity. *The European Journal of Finance*, 1–45. https://doi.org/10.1080/1351847X.2024.2433021

https://www.mevzuat.gov.tr/mevzuat?MevzuatNo=2872&MevzuatTur=1&MevzuatTertip=5

https://www.resmigazete.gov.tr/eskiler/2025/07/20250709-1.htm

Ikejiaku, B. V. (2014). International law is Western made global law: The perception of Third-World category. *African Journal of Legal Studies, 6*(2–3), 337–356.

Inderst, G., Kaminker, C., & Stewart, F. (2012). *Defining and measuring green investments: Implications for institutional investors' asset allocations* (OECD Working Papers on Finance, Insurance and Private Pensions, No. 24). OECD Publishing. https://doi.org/10.1787/5k9312twnn44-en

International Court of Justice. (1997). *Gabčíkovo-Nagymaros Project* (Hungary/Slovakia), Judgment (I.C.J. Reports 1997)

International Court of Justice. (2025, July 23). *Obligations of States in respect of climate change* (Advisory Opinion, Case No. 187) [PDF]. https://www.icj-cij.org/sites/default/files/case-related/187/187-20250723-adv-01-00-en.pdf

International Federation for Human Rights (FIDH) & Forum for Re-sponsible Investment (FIR). (2025, October 17). *Attacks against the Corporate Sustainability Due Diligence Directive: Industry pressure to repeal essential corporate accountability rules*. FIDH. Retrieved from https://www.fidh.org/en/issues/business-human-rights-environment/business-and-human-rights/attacks-against-csddd-fidh-fir-reaffirm-importance-of-due-diligence

International Finance Corporation. (2020). *Green Bond Handbook: A step-by-step guide to issuing a green bond* [PDF]. https://www.ifc.org/content/dam/ifc/doc/mgrt/202203-ifc-green-bond-handbook.pdf

International Institute for Sustainable Development. (n.d.). *What is sustainable development?* https://www.iisd.org/about-iisd/sustainable-development

International Law Commission. (2001). *Draft articles on responsibility of States for internationally wrongful acts, with commentaries*. Yearbook of the International Law Commission, 2001, Vol. II, Part Two.

Irawan, B., Firdaus, Jaya, B. P. M., Taufiqurrohman, A. H. A., Sari, S. W., & Furqoni, S. (2025). State responsibility and strategy in preventing and protecting Indonesian fisheries crews working on foreign fishing vessels from modern slavery. *Australian Journal of Maritime & Ocean Affairs, 17*(2), 186–206. https://doi.org/10.1080/18366503.2024.2333107

Kamu Gözetimi, Muhasebe ve Denetim Standartları Kurumu. (2023). *Türkiye Sürdürülebilirlik Raporlama Standartları 1 (TSRS 1): Sürdürülebilirlikle ilgili finansal bilgilerin açıklanmasına ilişkin genel hükümler* [Standard]. https://www.kgk.gov.tr/Portalv2Uploads/files/Duyurular/v2/Surdurulebilirlik/RaporlamaStandarti/TSRS%201.pdf?__blob=publicationFile&v=1

Katila, P., Colfer, C. J. P., de Jong, W., Galloway, G., Pacheco, P., & Winkel, G. (Eds.). (2019). *Sustainable development goals: Their impacts on forests and people*. Cambridge University Press.

Khot, U. A., Warchold, A., & Pradhan, P. (2026). Existing gaps in understanding Sustainable Development Goals interactions: Insights from a systematic review. *Environmental Impact Assessment Review, 118*, Article 108274. https://doi.org/10.1016/j.eiar.2025.108274

Kim, R. E. (2016). The nexus between international law and the sustainable development goals. *Review of European, Comparative and International Environmental Law*. https://doi.org/10.1111/reel.12148

Knaack, P., & Larionova, M. (2022). Global standard setting for sustainable finance: The role of the EU. *Journal of Environmental Investing, 13*(1), 1–18.

Koch, B. (2023). *The European Green Bond Standard: An in-depth analysis of its potential as the new benchmark in green bond markets*. https://doi.org/10.13140/RG.2.2.23536.43523

Lehmann, A. (2021, July 13). *The EU green bond standard: Sensible implementation could define a new asset class*. Bruegel.

Lima, D., & Pereira, C. M. (2024). European human rights law for a digital economy: A right to financial inclusion? *Revista Direito Mackenzie, 3*(18), e17517EN. https://doi.org/10.5935/2317-2622/direitomackenzie.v18n3.17517EN

Liu, S., & Zhang, H. (2024). Governance quality and green growth: New empirical evidence from BRICS. *Finance Research Letters, 65*, Article 105566.

Lukács B. (2025). Sovereign green bond reporting practices in the EU: A harmonised or fragmented landscape? *Cognitive Sustainability, 4*(4). https://doi.org/10.55343/CogSust.20624

Lukashuk, I. I. (1989). The principle Pacta Sunt Servanda and the nature of obligation under international law. *American Journal of International Law, 83*(3), 513–518. https://doi.org/10.2307/2203309

MacAskill, S., Roca, E., Liu, B., Stewart, R. A., & Sahin, O. (2021). Is there a green premium in the green bond market? Systematic literature review revealing premium determinants. *Journal of Cleaner Production, 280*, Article 124491. https://doi.org/10.1016/j.jclepro.2020.124491

Martinčević, I., Primorac, D., & Dorić, B. (2024). Corporate sustainability reporting directive (CSRD): Obligations, challenges and requirements for companies. *ENTRENOVA—ENTerprise REsearch InNOVAtion, 10*, 317–327. https://doi.org/10.54820/entrenova-2024-0026

Mieszkowska, J. (2024). *The unintended consequences of the EU Corporate Sustainability Due Diligence Directive* (American Journal of International Law). Cambridge University Press.

Mittwoch, A.-C., & Bremenkamp, F. L. (2023). The German supply chain act—A sustainable regulatory framework for internationally active market players? *Review of European and Comparative Law, 55*(4), 189–219. https://doi.org/10.31743/recl.16677

Monciardini, D. (2025). *Regulating EU sustainability reporting: Learning from failure and success*. Cambridge University Press.

Möslein, F. (2025). Legal foundations of Green FinTech: Paving the way for sustainable finance. In I. G. Hoven, S. Y. In, & T. Puschmann (Eds.), *Sustainable digital finance* (pp. 103–117). Springer. https://doi.org/10.1007/978-3-032-02983-6_7

Mubarik, M. S. (2025). Decarbonization through supply chain innovation: Role of supply chain collaboration and mapping. *Journal of Cleaner Production, 507*, 145492. Advance online publication. https://doi.org/10.1016/j.jclepro.2025.S095965262500842X

Naseem, M., & Alim, W. (2026). Fintech and sustainable development: Exploring the interactive role of governance. *Journal of Asian Development Studies, 15*(1), 59–72. https://doi.org/10.62345/

Neuberger, D. (2015). *Financial inclusion, regulation, and education in Germany* (ADBI Working Paper No. 530). Asian Development Bank Institute.

Ni, L., Li, L., Zhang, X., & Wen, H. (2022). Climate policy and foreign direct investment: Evidence from a quasi-experiment in Chinese cities. *Sustainability, 14*(24), 16469. https://doi.org/10.3390/su142416469

Ni, L., Yu, Y., & Wen, H. (2023, September 18). Impact of fintech and environmental regulation on green innovation: Inspiration from prefecture-level cities in China. *Frontiers in Ecology and Evolution, 11*, Article 1265531. https://doi.org/10.3389/fevo.2023.1265531

Oberthür, S., & Dupont, C. (2021). The European Union's international climate leadership: Towards a grand climate strategy? *Journal of European Public Policy, 28*(7), 1095–1114. https://doi.org/10.1080/13501763.2021.1918218

Och, M. (2025). The EU taxonomy regulation and the prevention of greenwashing. In K. Alexander, M. Gargantini, & M. Siri (Eds.), *The Cambridge handbook of EU sustainable finance: Regulation, supervision and governance* (pp. 357–380). Cambridge University Press. https://doi.org/10.1017/9781009483971.017

OECD. (2023). *Fintech and sustainable finance: Building the synergies.* https://www.oecd.org/finance/fintech/

Offiong, U. P., Szopik-Depczyńska, K., & Ioppolo, G. (2025). FinTech innovations for sustainable development: A comprehensive literature review and future directions. *Sustainable Development, 33*(S1), 1459–1471. https://doi.org/10.1002/sd.70068

Okeke, N., Bakare, O., & Achumie, G. (2024). Integrating policy incentives and risk management for effective green finance in emerging markets. *International Journal of Frontiers in Science and Technology Research, 7*(1), 76–88. https://doi.org/10.53294/ijfstr.2024.7.1.0050

Omeragić, N. (2023). *Fintech tools, regulations and limitations in achieving sustainable development goals* (Master's thesis, University of Sarajevo, School of Economics and Business).

Ortiz-Martínez, E., & Marín-Hernández, S. (2024). Sustainability information in European small- and medium-sized enterprises. *Journal of the Knowledge Economy, 15*, 7497–7522. https://doi.org/10.1007/s13132-023-01386-7

Özçelik, N., Rey García, M., & Mato Santiso, V. (2025). Enablers and barriers to European Green deal implementation: A systematic review and framework proposal. *Sustainable Production and Consumption.* https://doi.org/10.1016/j.spc.2025.10.018

Ozdamirova, L., Dzhamaldinova, M., & Samatova, C. (2024). The role of Legal mechanisms in the implementation of the Sustainable Development Goals (SDGs). *Reliability: Theory & Applications, 19*, 1359–1367.

Papari, C., Toxopeus, H., Polzin, F., Bulkeley, H., & Menguzzo, E. V. (2024). Can the EU taxonomy for sustainable activities help mobilise investments for the sustainability transition? *Environmental Science & Policy.* https://doi.org/10.1016/j.envsci.2023.103598

Patel, G. (2016). Human rights and the universal periodic review: Rituals and ritualism by Hilary Charlesworth and Emma Larking (Eds.): Cambridge: Cambridge University Press, 2014. *Human Rights Review, 17*, 405–407. https://doi.org/10.1007/s12142-016-0421-7

Philipp, G. (2024). Social rights. In R. Grote, F. Lachenmann,& R. Wolfrum (Eds.), *Max Planck encyclopedia of comparative constitutional Law* (law:mpeccol e172). Oxford University Press. https://doi.org/10.1093/law:mpeccol/law-mpeccol-e172

Purvis, B., Mao, Y., & Robinson, D. (2019). Three pillars of sustainability: In search of conceptual origins. *Sustainability Science, 14*(3), 681–695. https://doi.org/10.1007/s11625-018-0627-5

Puschmann, T., & Khmarskyi, V. (2024). Green fintech: Developing a research agenda. *Corporate Social Responsibility and Environmental Management, 31*(4), 2823–2837.

Pyka, M. (2023). The EU Green Bond Standard: A plausible response to the deficiencies of the EU green bond market? *European Business Organization Law Review, 24*(4), 623–643. https://doi.org/10.1007/s40804-023-00278-2

Qiao, C., Cai, W., & Chen, S. (2025). Green fintech contributes to environmental sustainability—Based on empirical evidence from China. *Humanities and Social Sciences Communications, 12*, 1895. https://doi.org/10.1057/s41599-025-06159-y

Radu, O. M., Dragomir, V. D., & Hao, N. (2023). Company level factors of non financial reporting quality under a mandatory regime: A systematic review of empirical evidence in the European Union. *Sustainability, 15*(23), 16265. https://doi.org/10.3390/su152316265

Republic of Türkiye. (2023, April 13). *Updated first nationally determined contribution (NDC)* [PDF]. https://unfccc.int/sites/default/files/NDC/2023-04/T%C3%9CRK%C4%B0YE_UPDATED%201st%20NDC_EN.pdf

Rossmann, F., Greitens, J., & Knoll, L. (2025). *SMEs within a data-driven sustainable finance framework: A European survey*. Centre for Financial Studies Working Paper Series.

Sachs, J. D. (2015). Achieving the sustainable development goals. *Journal of International Business Ethics, 8*(2), 53–62.

Sahintürk, Ö. (2025). EU legal framework on sustainable finance: Implications for non-EU financial market participants and extraterritorial compliance. *Stanford Journal of International Law*.

Saul, B., Kinley, D., & Mowbray, J. (Eds.). (2014). *The international covenant on economic, social and cultural rights: Cases, materials, and commentary*. Oxford University Press.

Savaresi, A. (2025). Verein Klima Seniorinnen Schweiz and Others v Switzerland: Making climate change litigation history: Review of European. *Comparative & International Environmental Law, 34*(1), 279–287. https://doi.org/10.1111/reel.12612

Schilling-Vacaflor, A. (2021). Putting the French duty of vigilance law in context: Towards corporate accountability for human rights violations in the Global South? *Human Rights Review, 22*(1), 109–127. https://doi.org/10.1007/s12142-020-00607-9

Schütze, F., & Stede, J. (2021). The EU sustainable finance taxonomy and its contribution to climate neutrality. *Intereconomics, 56*(4), 236–241. https://doi.org/10.1007/s10272-021-0993-2

Sharma, A., & Kukreja, S. (2013). An analytical study: Relevance of financial inclusion for developing nations. *Research Inventy: International Journal of Engineering and Science, 2*(6), 15–20.

Sharma, A., Adekunle, B. I., Ogeawuchi, J. C., Abayomi, A. A., & Onifade, O. (2021, March 31). Governance challenges in cross-border fintech operations: Policy, compliance, and cyber risk management in the digital age. *Iconic Research and Engineering Journals, 4*(9), 278–286. https://www.irejournals.com/formatedpaper/1708642.pdf

Sikora, A. (2020). European Green Deal—Legal and financial challenges of the climate change. *ERA Forum, 21*(4), 681–697. https://doi.org/10.1007/s12027-020-00637-3

Sinnig, J., & Zetzsche, D. A. (2025). The EU's corporate sustainability due diligence directive: From disclosure to mandatory prevention of adverse sustainability impacts in supply chains. *European Journal of Risk Regulation, 16*(2), 628–652. https://doi.org/10.1017/err.2024.100

Steuer, S., & Tröger, T. H. (2022, April). The role of disclosure in green finance. *Journal of Financial Regulation, 8*(1), 1–50. https://doi.org/10.1093/jfr/fjac001

Syed, R. F., & Ikra, M. (2023). Industrial killing in Bangladesh: State policies, common-law nexus, and international obligations. *Employee Responsibilities & Rights Journal, 35*(4), 601–618. https://doi.org/10.1007/s10672-022-09431-4

T.C. Cumhurbaşkanlığı Finans Ofisi. (2025, February 17). *Türkiye Fintek Ekosistemi Durum Raporu 2024*.

T.C. Cumhurbaşkanlığı Strateji ve Bütçe Başkanlığı. (2023, December 11). *On İkinci Kalkınma Planı (2024–2028)*. https://www.sbb.gov.tr/wp-content/uploads/2023/12/On-Ikinci-Kalkinma-Plani_2024-2028_11122023.pdf

Tams, C. J. (2018). Decisions of the International Court of Justice as sources of international law? In *Decisions of the ICJ as sources of international law?* (pp. 21–38). Gaetano Morelli Lectures Series, Vol. 2. International and European Papers Publishing. https://crde.europeanpapers.eu/sites/default/files/GMLS%20-%20Decisions%20of%20the%20ICJ%20as%20Sources%20of%20International%20Law%20(2018).pdf

Tanç, Ş. G., & Çardak, D. (2024). Borsa İstanbul Sürdürülebilirlik 25 Endeksindeki işletmelerin sürdürülebilirlik performanslarının TOPSIS yöntemi ile sıralanması. *Hitit Ekonomi Ve Politika Dergisi, 4*(2), 107–125.

Thirlway, H. (2019). *The sources of international law* (2nd ed.). Oxford University Press.

Thorens, M., Bernaz, N., & Hospes, O. (2025). Advocating for the EU Corporate Sustainability Due Diligence Directive against the odds: Strategies and legitimation. JCMS: *Journal of Common Market Studies, 63*(2), 606–623. https://doi.org/10.1111/jcms.13623

United Nations. (1966, December 16). *International Covenant on Economic, Social and Cultural Rights* [Treaty; UN TS 993 (16)]. https://treaties.un.org/pages/ViewDetails.aspx?chapter=4&mtdsg_no=IV-3&src=IND

United Nations. (2015, December 12). *Paris Agreement*. United Nations Treaty Collection. https://treaties.un.org/pages/viewdetails.aspx?src=treaty&mtdsg_no=xxvii-7-d&chapter=27&clang=_en

United Nations. (n.d.). *Millennium Development Goals (MDGs) and development agenda (retired)*. United Nations Sustainable Development. https://www.un.org/sustainabledevelopment/development-agenda-retired/

United Nations. (n.d.). *Sustainable development goals*. https://sdgs.un.org/goals

United Nations Department of Public Information. (2017). *Sustainable development goals*. Stylus Publishing.

United Nations Environment Programme Finance Initiative. (n.d.). *Positive impact* [Web page]. UNEP FI. Retrieved October 15, 2025, from https://www.unepfi.org/positive-impact-2/

United Nations General Assembly. (2015). *Transforming our world: The 2030 Agenda for Sustainable Development* (Res. A/RES/70/1). United Nations.

van den Hurk, A., & van der Klooster, I. (2024). The European Commission's sustainable finance action plan and other international initiatives. In D. Busch, G. Ferrarini, & S. Grünewald (Eds.), *Sustainable Finance in Europe*. EBI Studies in Banking and Capital Markets Law. Palgrave Macmillan, Cham. https://doi.org/10.1007/978-3-031-53696-0_2

Verhoeven, J. (2002). Considération sur ce qui est commun: Cours général de droit international public. *Recueil des Cours De L'académie De Droit International De La Haye, 334*, 9–112.

Wang, K.-H., Zhao, Y.-X., Jiang, C.-F., & Li, Z.-Z. (2022). Does green finance inspire sustainable development? Evidence from a global perspective. *Economic Analysis and Policy, 75*, 412–426. https://doi.org/10.1016/j.eap.2022.06.002

Williment, C. (2025, October 13). Why are Siemens & TotalEnergies CEOs against the EU's Omnibus? *Sustainability Magazine*. https://www.sustainabilitymag.com/news/why-are-46-ceos-against-the-eus-sustainability-omnibus/

Wilson, J., Adeyemi, T., & Rahman, N. (2025). *Hybrid governance models: Integrating formal and informal institutions for environmental sustainability*. https://www.researchgate.net/publication/396374100_Hybrid_Governance_Models_Integrating_Formal_and_Informal_Institutions_for_Environmental_Sustainability

World Bank. (n.d.). *Financial inclusion overview*. https://www.worldbank.org/en/topic/financialinclusion/overview

World Bank. (n.d.). *World Bank climate change action plan*. https://www.worldbank.org/en/topic/climatechange/brief/world-bank-group-climate-change-action-plan

World Bank Group. (2021). World Bank Group Climate Change Action Plan 2021–2025: Supporting Green, Resilient, and Inclusive Development., *World Bank Group*, https://openknowledge.worldbank.org/server/api/core/bitstreams/19f8b285-7c5b-5312-8acd-d9628bac9e8e/content

World Meteorological Organization. (2025, January 10). *WMO confirms 2024 as warmest year on record, at about 1.55 °C above pre-industrial level* [Press release]. https://wmo.int/news/media-centre/wmo-confirms-2024-warmest-year-record-about-155degc-above-pre-industrial-level

Wrede, P. F. W. (2018, July). *How technology can make insurance more inclusive (FinTech Note No. 2)*. World Bank. https://documents1.worldbank.org/curated/en/583381531209953337/pdf/128157-9-7-2018-11-49-10-FintechNotesTechnologyInsuranceInclusiveFinalLowRes.pdf

Wu, Q. (2024). From bits to emissions: How FinTech benefits climate resilience? *Empirical Economics, 67*(5), 2009–2037. https://doi.org/10.1007/s00181-024-02609-9

Xu, J., Li, T., Chen, F., & Quan, Y. (2025). Does Fintech foster the development of green finance? *Emerging Markets Finance & Trade, 61*(6), 1771–1781. https://doi.org/10.1080/1540496X.2024.2430500

Yoshino, N., & Morgan, P. J. (2016). *Overview of financial inclusion, regulation, and education* (ADBI Working Paper No. 591). Asian Development Bank Institute.

Zarei, S., Poorhashemi, S. A., & Pournouri, M. (2015). A deliberation on recent legal cases and judgments of the International Court of Justice on environmental disputes. *International Journal of Environmental Science and Technology, 12*(10), 3391–3404. https://doi.org/10.1007/s13762-015-0825-y

Zetzsche, D. A., Bodellini, M., & Consiglio, R. (2022). The EU sustainable finance framework in light of international standards. *Journal of International Economic Law, 25*(4), 659–679. https://doi.org/10.1093/jiel/jgac043

Zimmermann, R. (2016, March 07). *Obligation.* Oxford Classical Dictionary. Retrieved 13 Oct. 2025, from https://doi.org/10.1093/acrefore/9780199381135.001.0001/acrefore-9780199381135-e-4492

Zoting, S. (2025). *Insurtech market size to hit USD 609.50 billion by 2034*. Precedence Research. Retrieved October 17, 2025, from https://www.precedenceresearch.com/insurtech-market

Conclusion

Barış C. Cantürk

Fintech and sustainable development play a profound role in the global financial ecosystem, which produces intertwined impacts and outcomes within the realm of law and regulation. For that reason, this study examines the subject from this perspective.

Each chapter, which endeavors to articulate distinct areas of expertise within the subject, could be subjected to a highly detailed summary. Indeed, even if such an exhaustive analysis were restricted solely to the outcomes and conclusions, it would span pages equivalent in length to the chapters themselves. Moreover, it would require reiterating each point of the individual outcomes. However, this exceeds the limit and the aim of this section. Therefore, the following paragraphs aim merely to reiterate the most critical points of the said chapters.

In this regard, the first chapter presents a review of the technical literature on fintech and sustainable development, with a particular emphasis on their intersection with regulatory frameworks.

The following chapter focuses on the relationship between fintech and Sustainable Development Goals (SDGs). In this context, it should be emphasized that the correlation between these two innovative concepts is multifaceted and transformative. To articulate this more explicitly, fintech promotes access to financial services, entrepreneurship, and investments via "SDG-friendly" fintech instruments, namely green financial tools, alternative and digital finance, and low-carbon activities. In addition to that, it should be underlined that its full potential requires substantial investments, especially in digital infrastructure and human capital, while supported through a regulatory framework. On the other hand, several risks (e.g., ethical concerns, cybersecurity, data law issues, and greenwashing) underscore the need for collaborative mechanisms and efforts across sectors. To summarize, as a powerful

B. C. Cantürk (✉)
Department of Information and Technology Law, Faculty of Law, Marmara University, Istanbul, Türkiye
e-mail: b.canturk@marmara.edu.tr

Future Finance Law Hub, Istanbul, Türkiye

B. Ş. Şahin and B. C. Cantürk (eds.), *Fintech and Sustainable Development*, Sustainable Finance, https://doi.org/10.1007/978-3-032-31020-0_8

means for SDGs, fintech encapsulates, on the one hand, opportunities and, on the other hand, risks and responsibilities. Therefore, it necessitates strong collaboration to foster safeguards and mitigate potential vulnerabilities.

Regulations constitute one of the foundational pillars of this domain. A primary deficiency in this field is the lack of legal certainty and clarity. In particular, the rapid pace of technological advancement makes it exceedingly difficult for the law to keep pace. This regulatory lag is most acutely observed in sectors such as crypto-assets and AI-driven fintech. To keep pace with these developments, several jurisdictions are undertaking significant measures; this chapter exhaustively details many of these efforts, with a primary focus on the European Union (EU) and Türkiye. Notably, legislative frameworks enacted within the EU, such as the Markets in Crypto-Assets Regulation (MiCA), the Artificial Intelligence Act, and the General Data Protection Regulation (GDPR), serve as prime exemplars of this regulatory awareness. Nevertheless, overregulation and its side effects on innovation, particularly within the EU are also emphasized.

The subsequent chapter conducts a highly detailed, subject-specific evaluation and comparative analysis, building upon the regulation list mentioned in the previous chapter. Namely, this chapter compares the regulations in Türkiye and the EU, especially regarding e-money, digital banks, open banking, and the crowdfunding sector. By doing so, the chapter aims to elucidate the impact of fintech regulations on sustainable development. All of these examinations indicate that the regulatory approaches of Türkiye and the EU are closely aligned. One of the fundamental reasons for this convergence is the parallel trajectory of fintech development across both jurisdictions, which has led to the availability of similar products and services. Although certain discrepancies exist between the two legal systems, the fundamental approach in both is to support fintech enterprises, which present a significant opportunity for advancing financial inclusion, while simultaneously establishing a reliable system to safeguard the data and assets of fintech users.

It is undeniable that businesses hold vital responsibilities to achieve progress on the SDGs; via fintech, it assigns crucial responsibilities to corporate entities as well as to public authorities. Indeed, while this constitutes an obligation for businesses, it simultaneously yields significant benefits for the entities themselves. In this chapter, which revolves around this subject, this dynamic gives rise to the concept of "responsible business". Responsible businesses distinguish themselves not merely through the traditional corporate paradigm of shareholder value maximization, but also through the wealth they generate for broader stakeholders and the negative externalities they voluntarily mitigate. In this context, the relevant subchapter examines the effects and role of the capital asset pricing model on responsible business. Following that, soft law on the importance of the SDGs and their impact on responsible practices has been revealed. Then, a potential leading catalyst, "sustainability-oriented innovation" is positioned as a factor to generate value for both shareholders and the other stakeholders. Lastly, this chapter posits that fintech plays a catalytic role in the SDGs, innovation, and responsible business, which are fundamentally interdependent.

The final chapter on niche issues at that intersection examines governmental responsibilities and, therefore, actions on SDGs via fintech. As regards the role of SDGs in international consensus regarding social, economic, and environmental issues, it is posited that the fintech ecosystem, with its innovative means, may strongly support achieving those objectives. However, the lack of international cooperation on regulations should be considered a primary hurdle.

MIX
Papier aus verantwortungsvollen Quellen
Paper from responsible sources
FSC® C105338

If you have any concerns about our products,
you can contact us on
ProductSafety@springernature.com

In case Publisher is established outside the EU,
the EU authorized representative is:
Springer Nature Customer Service Center GmbH
Europaplatz 3, 69115 Heidelberg, Germany

Printed by Libri Plureos GmbH
in Hamburg, Germany